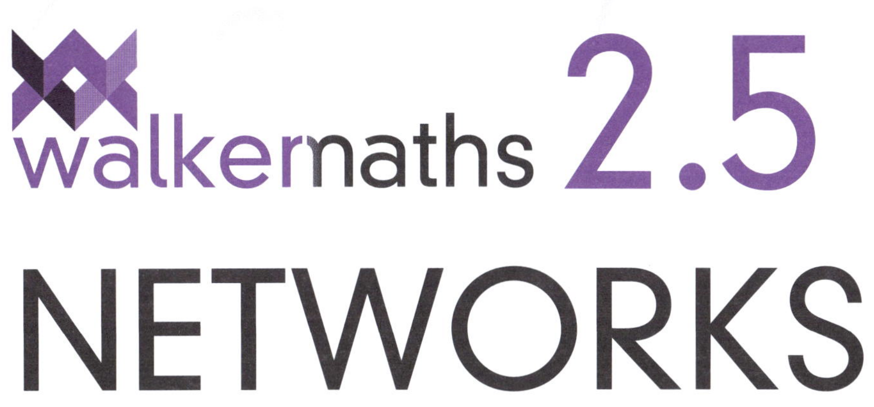

NCEA Level 2 Internal

Charlotte Walker and Victoria Walker

Walker Maths 2.5 Networks
1st Edition
Charlotte Walker
Victoria Walker

Editor: Eva Chan
Designer: Cheryl Smith, Macarn Design
Production controller: Siew Han Ong

Acknowledgements
Cover photo courtesy of Shutterstock.

We wish to thank the Boards of Trustees of Darfield and Riccarton High Schools for allowing us to use materials and ideas developed while teaching. Our thanks also go to all past and present colleagues who have generously shared their expertise and ideas.

For product information and technology assistance,
in Australia call **1300 790 853**;
in New Zealand call **0800 449 725**

For permission to use material from this text or product, please email **aust.permissions@cengage.com**

National Library of New Zealand Cataloguing-in-Publication Data
A catalogue record for this book is available from the National Library of New Zealand.

978 0 17 038943 3

Cengage Learning Australia
Level 7, 80 Dorcas Street
South Melbourne, Victoria Australia 3205

For learning solutions, visit **cengage.co.nz**

Printed in China by 1010 Printing International Ltd
16 26

CONTENTS

Glossary 4

What are networks? 5

Nodes and edges 6

The degree of a node 8

Odd and even nodes 9

Paths and circuits 11

Constructing networks 13

Shortest paths 20

Traversability 27

Putting it all together 32

Networks from tables 36

Spanning trees 43
- Minimum spanning trees 43
- Maximum spanning trees 47

Practice tasks 50

Answers 62

ISBN: 9780170389433

Glossary

Make your own glossary of key terms:

Term	Definition	Picture/Example/Other names
Network		
Node		
Vertex (plural 'vertices')		
Edges		
Weighted path		
Shortest path		
Traversability		
Minimum spanning tree		
Maximum spanning tree		
Euler circuit		
Euler path		

ISBN: 9780170389433

What are networks?

Networks are systems for interconnecting people or things.

Examples: Road and rail maps, wiring diagrams for a house, orienteering courses, nerve networks in the body, computer networks, courier routes, sewers, etc.

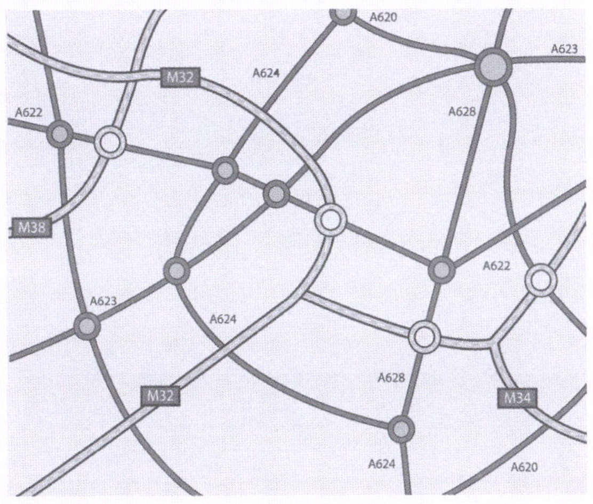

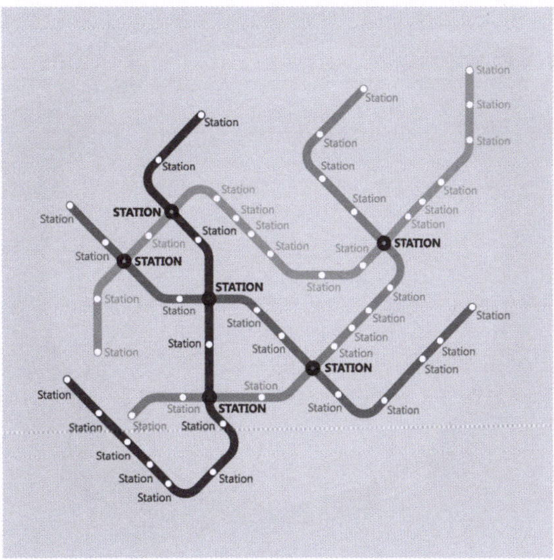

You need to be able to:

- identify shortest paths
- understand and apply the concept of traversability
- use information given in a table to produce a network
- find minimum and maximum spanning trees.

ISBN: 9780170389433

Nodes and edges

- Networks are composed of **nodes** and **edges**.
- Nodes are also called vertices (one vertex, several vertices).
- Edges are also called arcs.

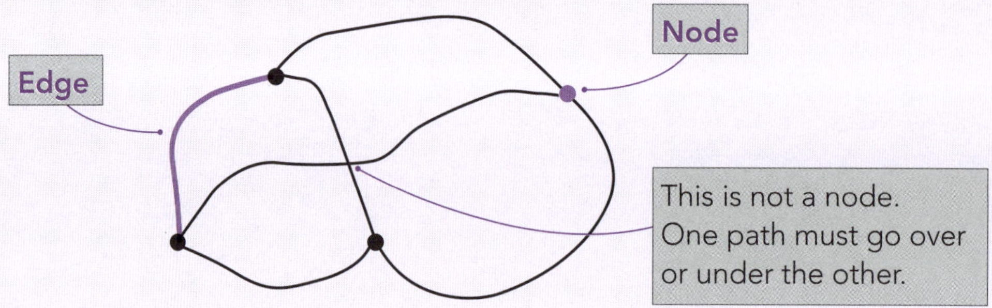

Practical contexts:

Context	Nodes	Edges
Road maps	Towns	Roads
Wiring of a house	Appliances	Wires
Nerve networks	Eyes, ears, tongue, etc.	Nerves
Communication systems	Phones, computers, etc.	Wires or Wi-Fi connections
Courier routes	Drop-off points	Roads

Examples: How many nodes and edges do the following networks have?

1

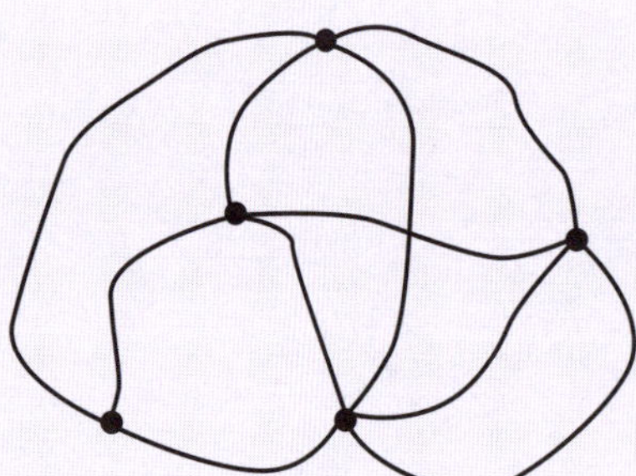

Nodes: 5

Edges: 10

2

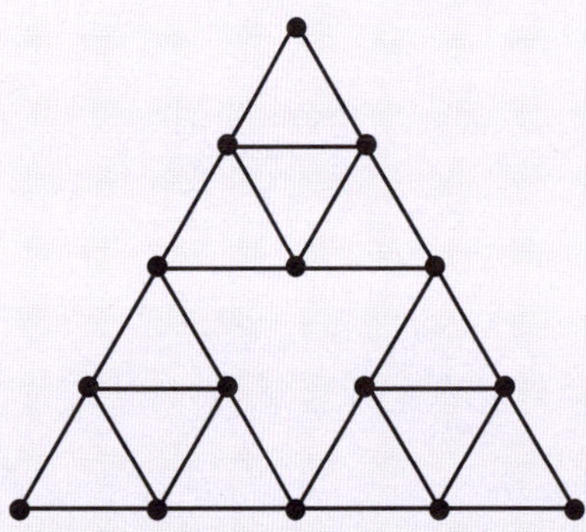

Nodes: 15

Edges: 27

ISBN: 9780170389433

Write down the numbers of nodes and edges in each of these networks.

1

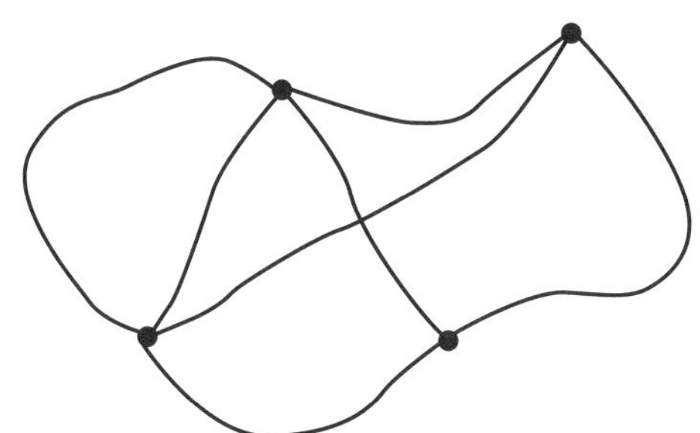

Nodes: ____________

Edges: ____________

2

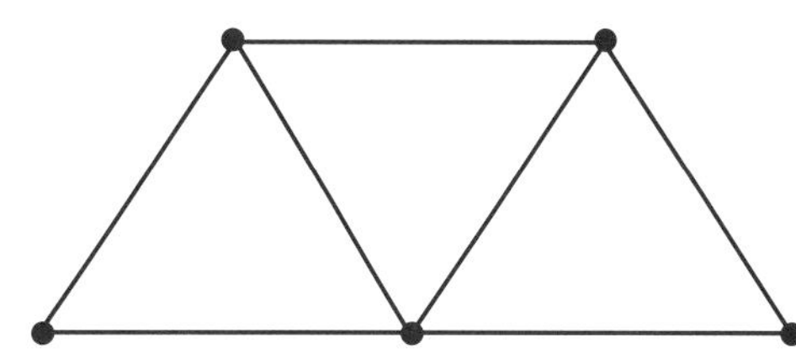

Nodes: ____________

Edges: ____________

3

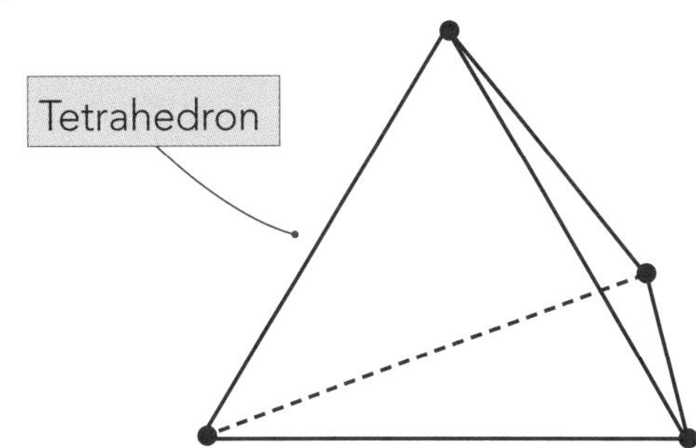

Nodes: ____________

Edges: ____________

4

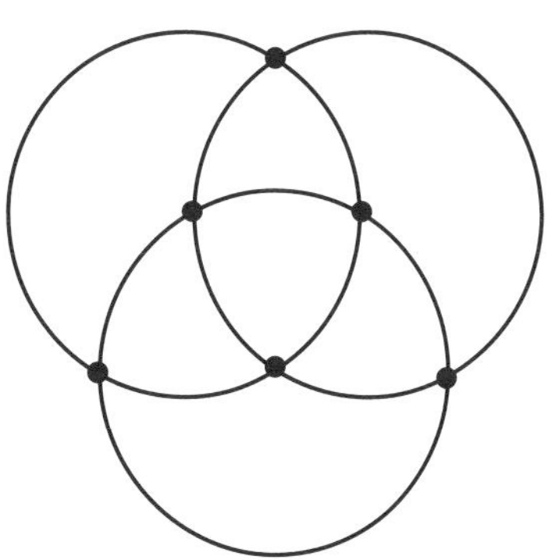

Nodes: ____________

Edges: ____________

5

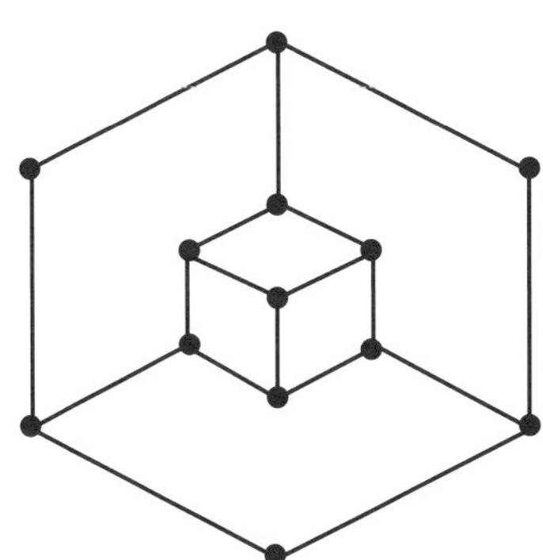

Nodes: ____________

Edges: ____________

6

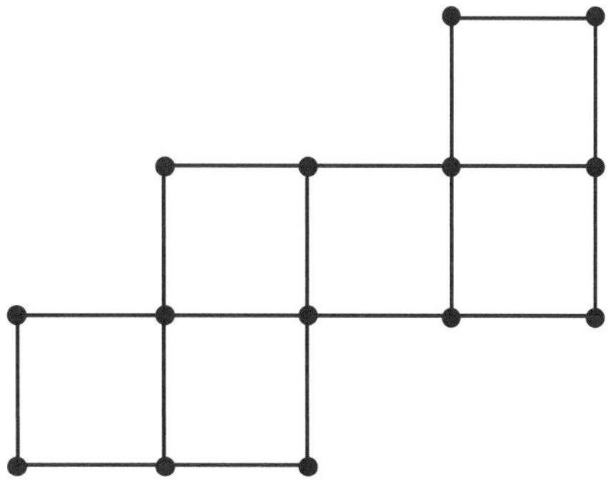

Nodes: ____________

Edges: ____________

ISBN: 9780170389433

The degree of a node

The **degree** of a node = the **number of edges** leading from it. For example, the number of roads leading out of a town would be the degree of the node.

Example:

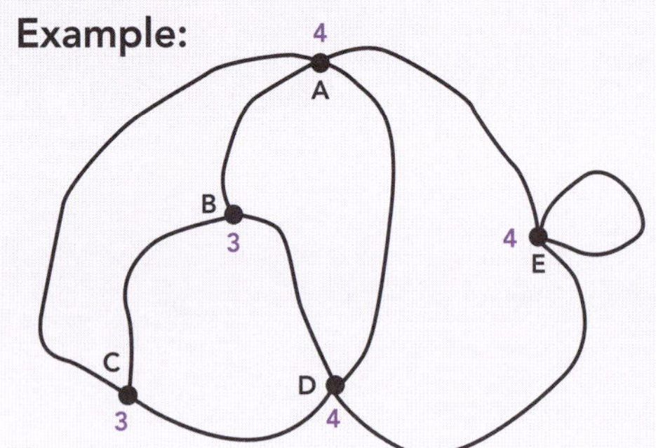

You may find drawing a table useful:

Node	A	B	C	D	E
Degree	4	3	3	4	4

Write down the degree of each node in the following networks.

1

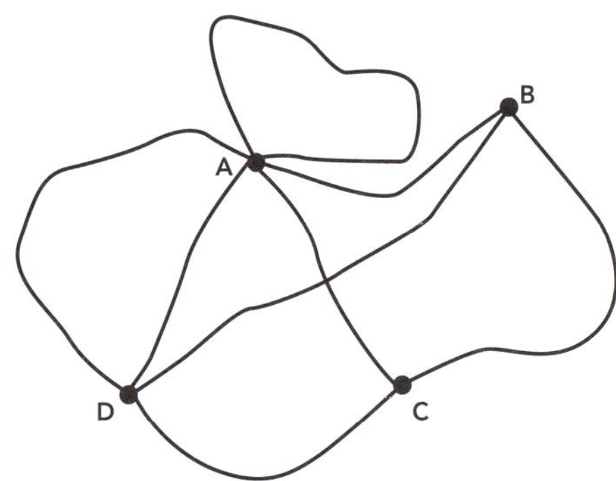

Node	A	B	C	D
Degree				

2

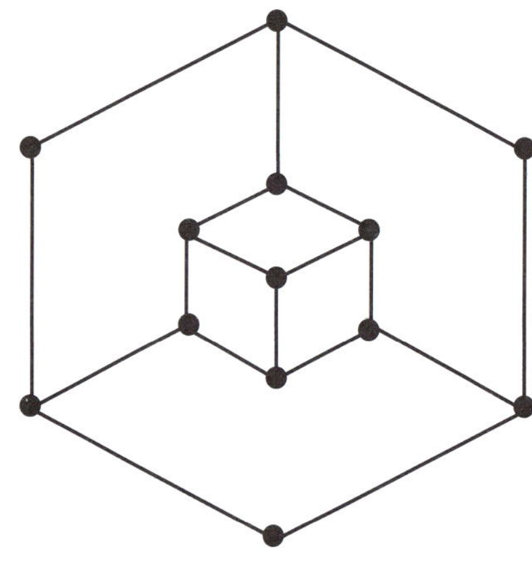

3

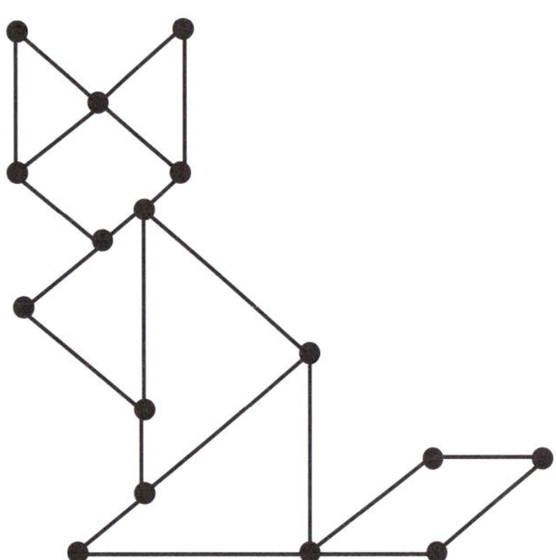

4

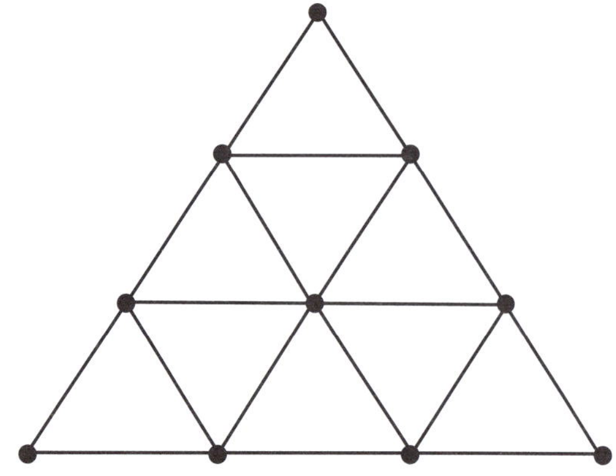

ISBN: 9780170389433

Odd and even nodes

- **Odd** node ⇒ the number of edges leading to or from the node is **odd**.
- **Even** node ⇒ the number of edges leading to or from the node is **even**.

Example:

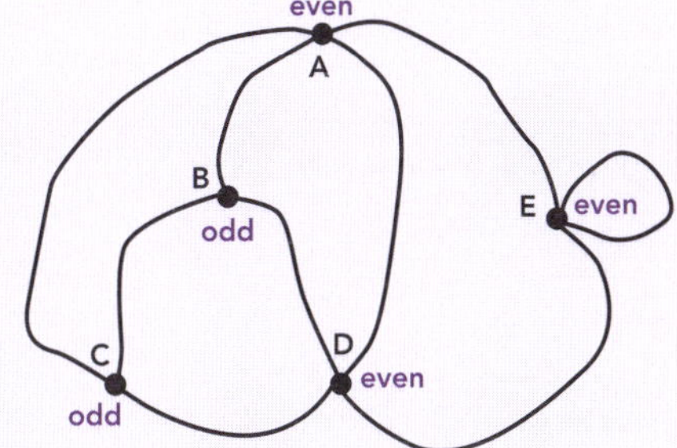

You may find drawing a table useful:

Node	A	B	C	D	E
Degree	4	3	3	4	4
Odd/Even	Even	Odd	Odd	Even	Even

Label each node in the following networks with 'odd' or 'even', and complete the tables.

1

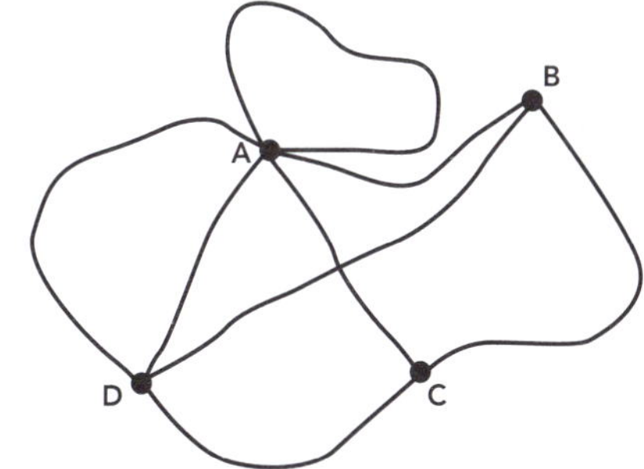

Node	A	B	C	D
Degree				
Odd/Even				

2

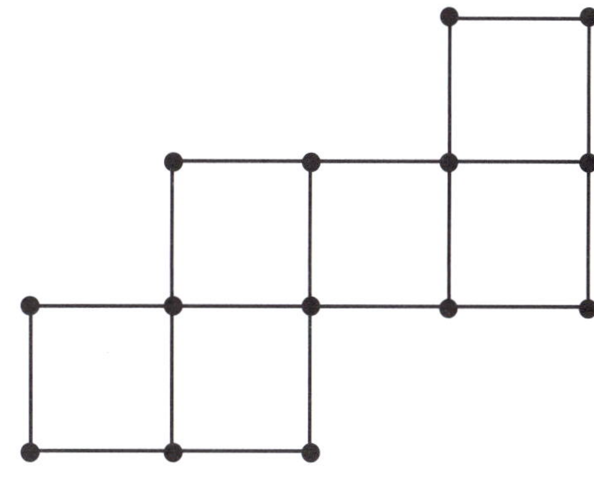

3

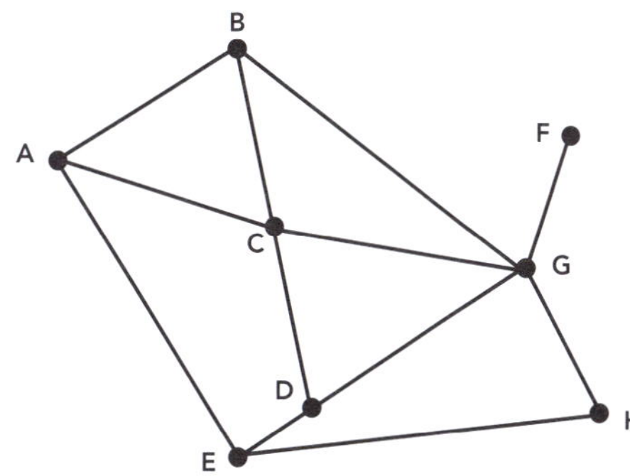

Node	A	B	C	D	E	F	G	H
Degree								
Odd/Even								

ISBN: 9780170389433

4

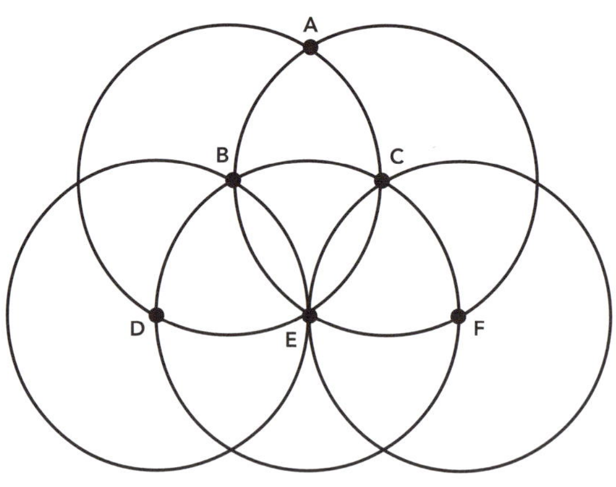

Node	A	B	C	D	E	F
Degree						
Odd/Even						

5

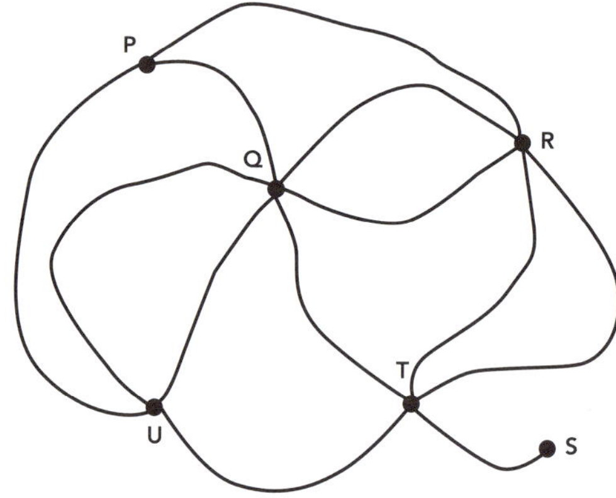

Node	P	Q	R	S	T	U
Degree						
Odd/Even						

6

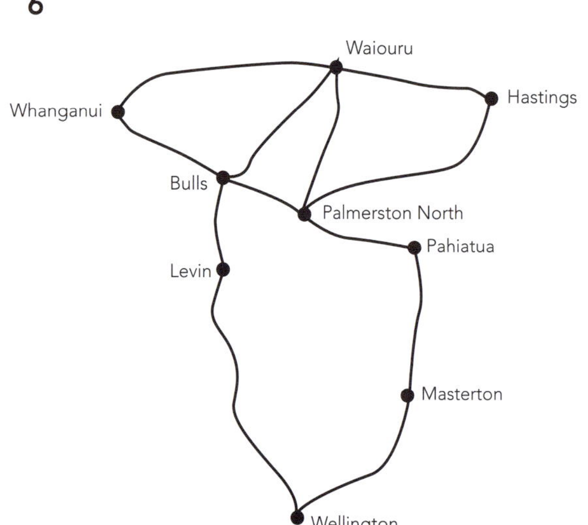

Node	Degree	Odd/Even
Whanganui		
Bulls		
Waiouru		
Hastings		
Palmerston North		
Pahiatua		
Levin		
Wellington		
Masterton		

ISBN: 9780170389433

Paths and circuits

A **path** starts and finishes at **different** points.

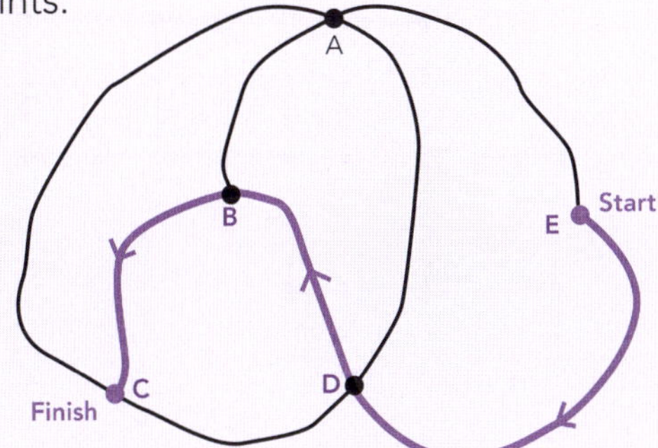

This path could be written as
E → D → B → C

A **circuit** starts and finishes at the **same** point.

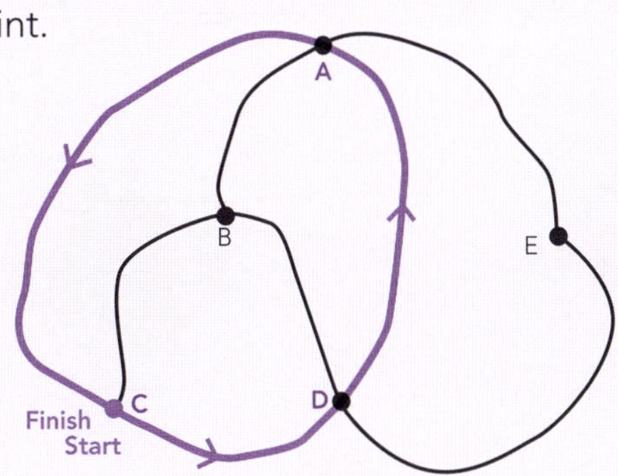

This circuit could be written as
C → D → A → C

Weighted paths and circuits

- Weighted paths and circuits have a **value** placed on each edge.
- The weightings could be distance, time, cost, etc.
- The value of a path or circuit is found by **adding** all of its edges.

Examples:

1

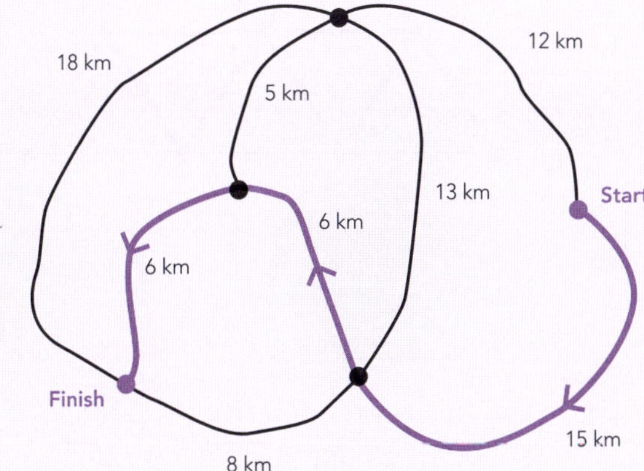

This **path** has a value of
15 + 6 + 6 = **27 km**.

2

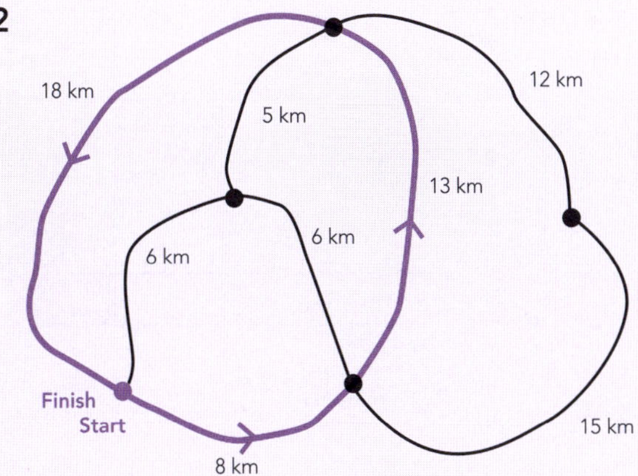

This **circuit** has a value of
8 + 13 + 18 = **39 km**.

ISBN: 9780170389433

State whether the following are paths or circuits, and give their weighted value.

1

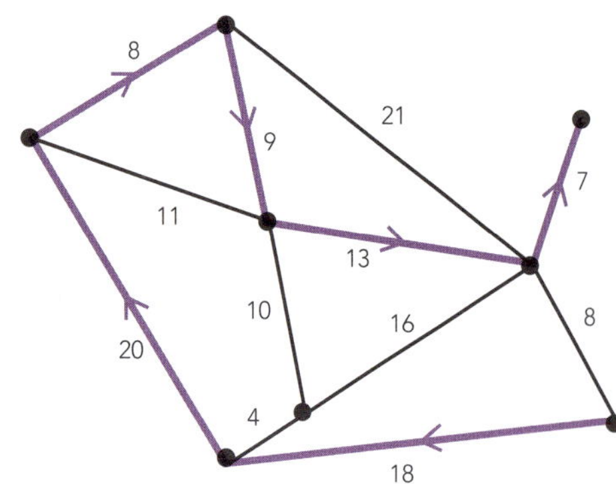

Path/Circuit: ____________

Value: ____________

2 The horizontal or vertical distance between adjacent dots is 1 unit.

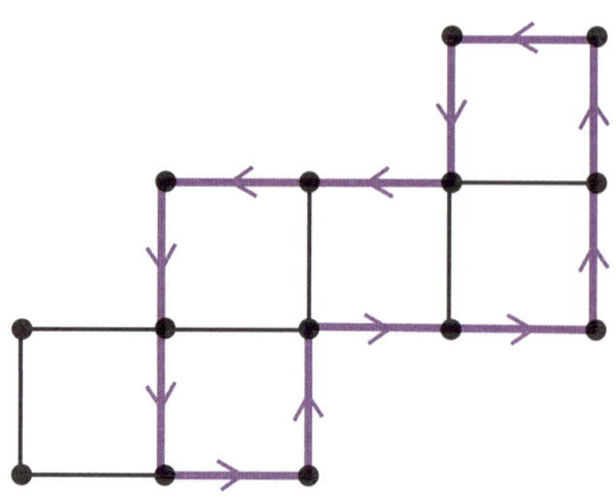

Path/Circuit: ____________

Value: ____________

3

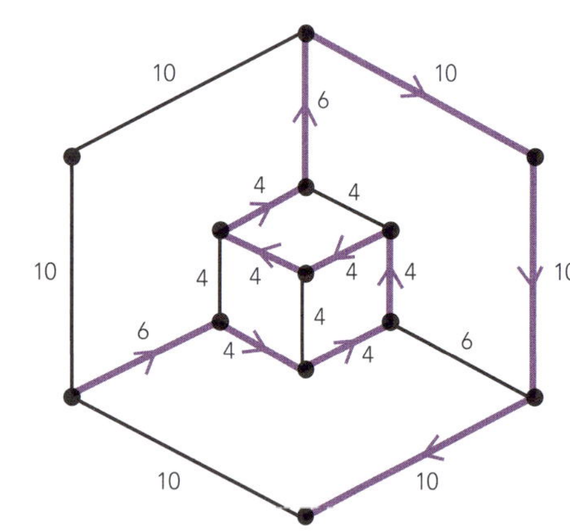

Path/Circuit: ____________

Value: ____________

4 The diagonal distance between adjacent dots is 1 unit.

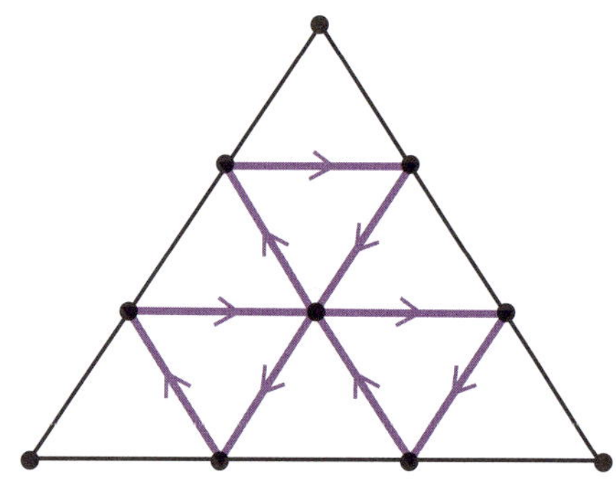

Path/Circuit: ____________

Value: ____________

5 The tourist value (1 – 10) of each road is shown.

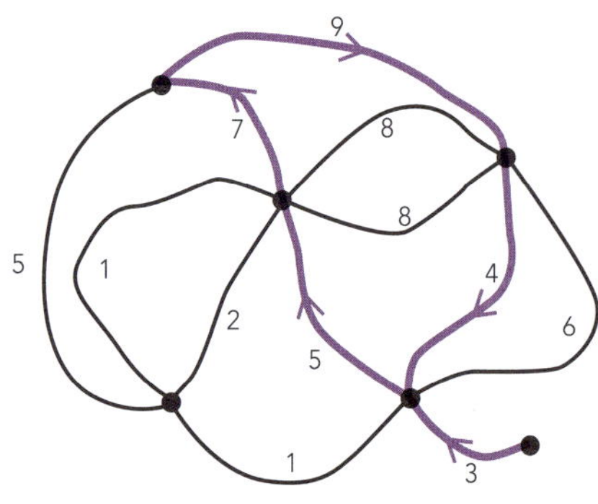

Path/Circuit: ____________

Value: ____________

6 The horizontal or vertical distance between adjacent dots is 10 units.

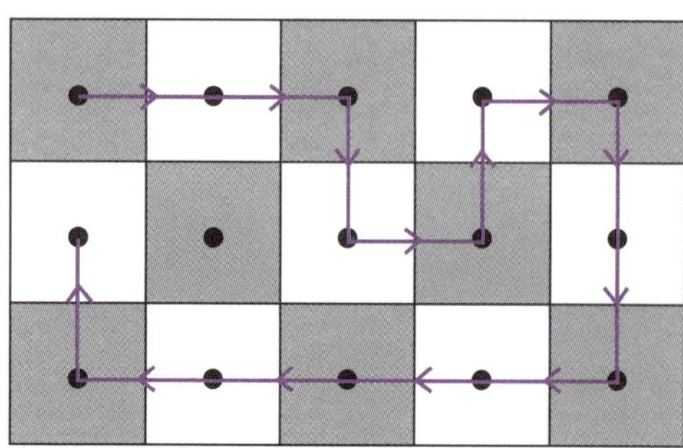

Path/Circuit: ____________

Value: ____________

ISBN: 9780170389433

Constructing networks

- Sometimes you need to convert information into network diagrams.
- Make the points that need to be connected nodes, and then use edges for the connections.

Example 1: Construct a network for the following: Aroha knows Maia, Toby, Ned and Chris; Chris and Maia both know Toby, but they don't know each other; Liz knows only Toby and Nick.

Step 1: Since Aroha knows everybody, draw her node at the centre, and the others around it.

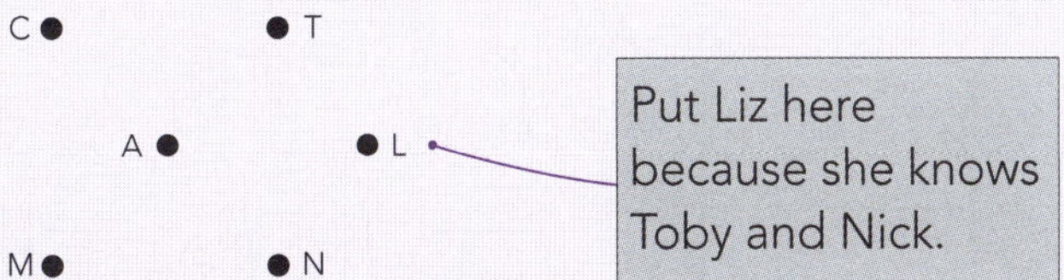

Step 2: Add edges for each connection from Maia.

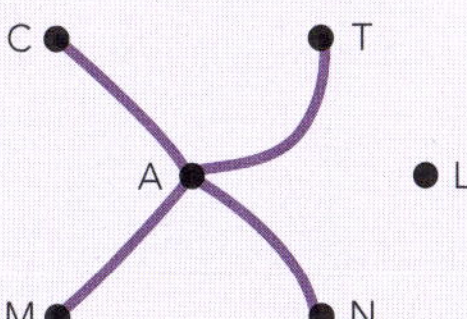

Step 3: Add edges for connections for Chris and Maia.

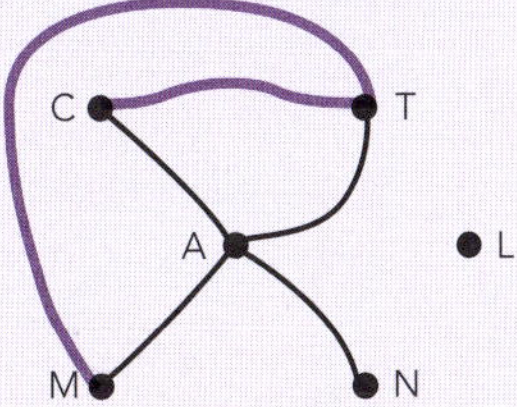

Step 4: Add connections from Liz.

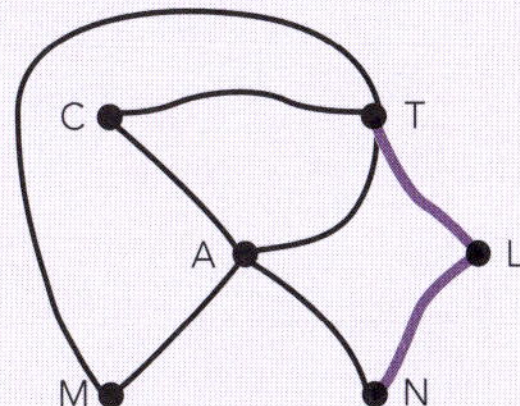

Note: One lot of information can sometimes be drawn in several ways:

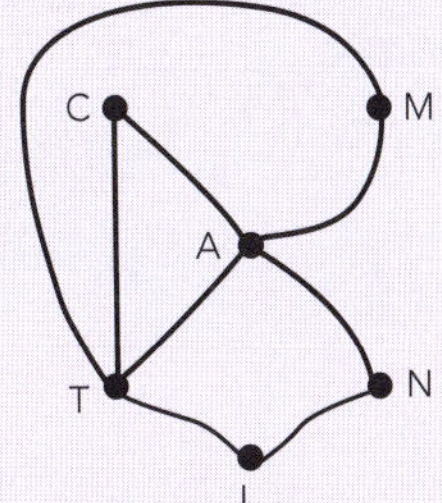

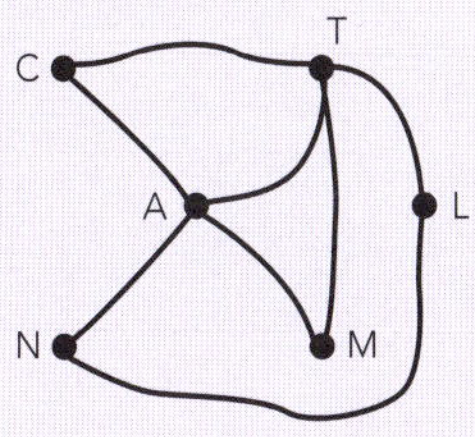

ISBN: 9780170389433

Example 2: Construct a network for the house plan shown below.

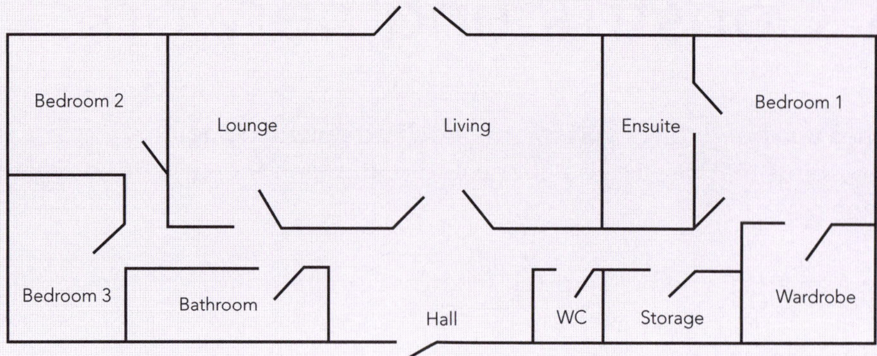

Step 1: Place a **node** in every room (including the hall).

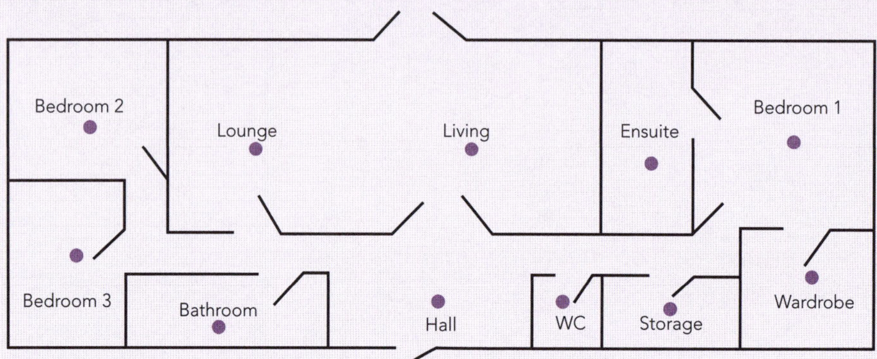

Step 2: Place an **edge** between the nodes wherever there is a doorway.

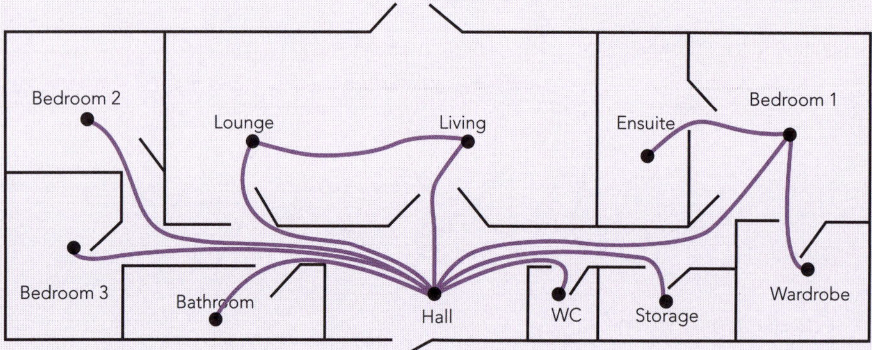

Step 3: Draw the network without the plan.

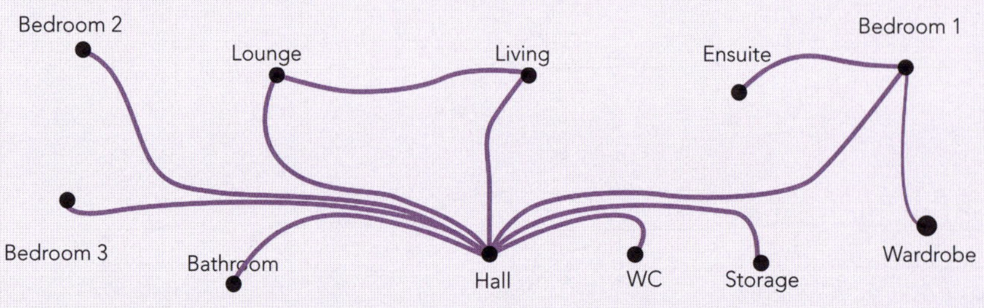

ISBN: 9780170389433

Where possible, draw networks to represent the following.

1 **a** A network with 5 nodes and 8 edges.

b A network with 4 nodes and 9 edges.

c A network with 2 nodes and 4 edges.

d A network with 3 nodes and 5 edges.

e A network with 2 odd nodes and 1 even node.

f A network with 2 even nodes and 1 odd node.

g A network with 2 odd nodes and 2 even nodes.

h A network with 3 odd nodes and 1 even node.

i A network with 3 odd nodes and 2 even nodes.

j A network with 2 odd nodes and 3 even nodes.

2 Maddy, Seone and Hana are friends on Facebook. Lisa is friends with Seone and Hana, but not Maddy. Jack is friends with only Hana and Seone.

3 Rogue Rentals has dealings with Trusty Tyres, Mick's Mechanics, and Fred's Financing. Trusty Tyres also has dealings with Fred's Financing, Dodgy Dealers, and Crafty Cars. Crafty Cars also deals with Fred's Financing and Mick's Mechanics. Mick's Mechanics also deals with Fred's Financing.

ISBN: 9780170389433

4 Consider each room (including the hall) to be a node, and draw edges between rooms where there is an interconnecting doorway.

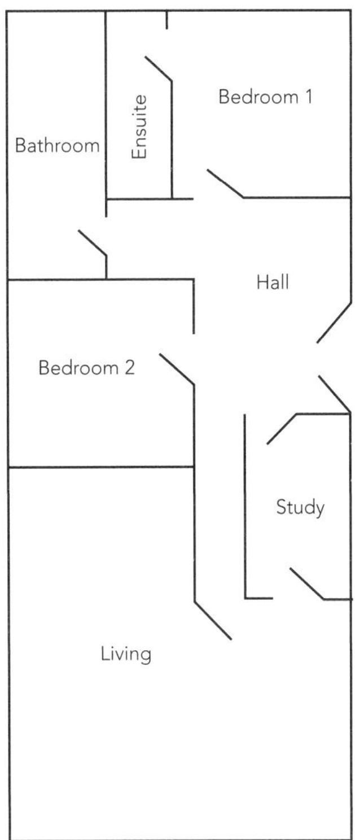

5 Consider each state/territory (not including Tasmania and ACT) to be a node, and construct edges between those that have a common boundary.

6 The diagram shows areas on a game board. Consider each area to be a node, and draw edges between adjacent areas.

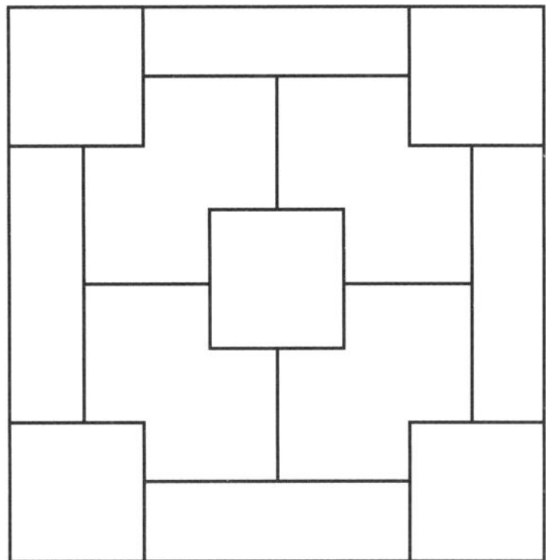

7 The diagram shows a small farm. Consider each area (House, Sheds, paddocks and Laneway) to be nodes. Construct edges between nodes where there are gates between adjacent areas.

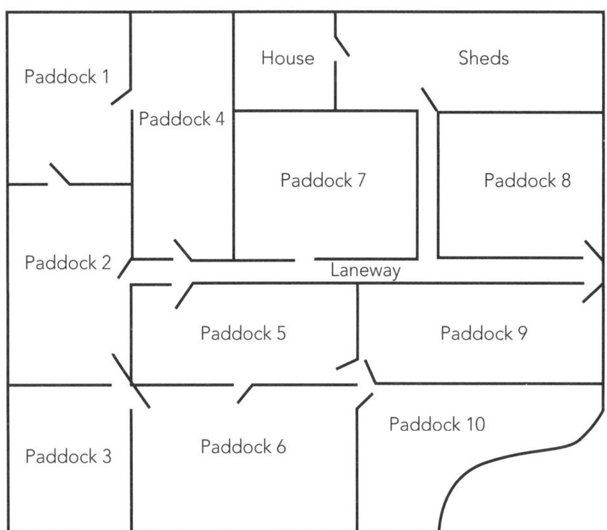

8 There are roads that connect the following places:

- Murchison and Nelson
- Springs Junction and Inangahua
- Reefton and Inangahua
- Springs Junction and Greymouth
- Westport and Inangahua
- Greymouth and Karamea
- Springs Junction and Christchurch
- Picton and Nelson.

Construct a network showing these centres as nodes, with the roads as edges that connect them.

ISBN: 9780170389433

9 For each island (not including Stewart Island), construct a network with each region as a node, and edges between regions with a common boundary. Then shade the **odd** regions of New Zealand.

North Island

South Island

Shortest paths

Shortest paths

- These are the shortest path from one point to another.
- They do **not** have to pass through every node.

Most common uses:

- In mapping systems to identify the quickest/cheapest/shortest route from one place to another.

Note: 1 There can be more than one path with the same shortest value.
2 Other factors may need to be considered if you are choosing the 'best' option.

Ways of finding the shortest paths:

1 By reasoning — for simpler networks.
2 Using a tree diagram — for more complex networks.

Using a tree diagram

Example: Find the shortest route(s) from the start (A) to the finish (H). The distances are in metres.

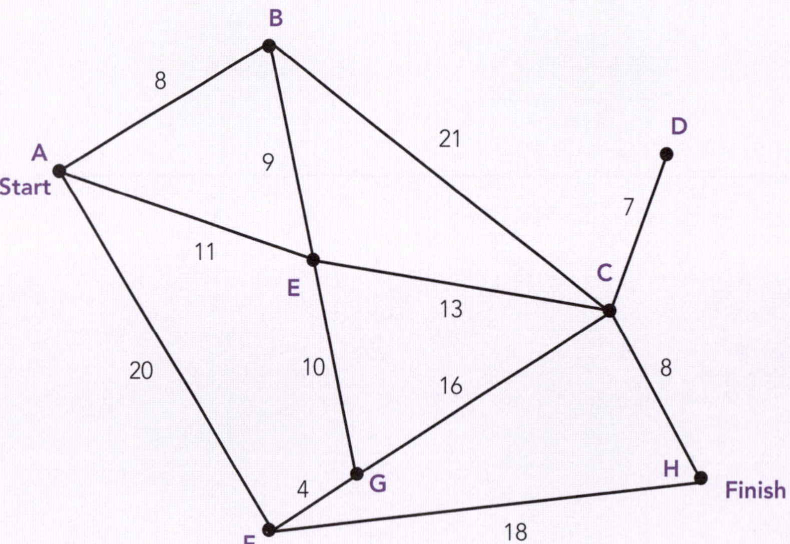

To create a tree diagram

1 Begin at the starting point and create a branch for each edge.

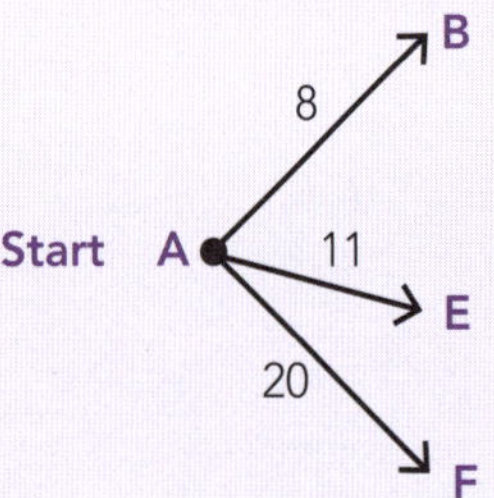

ISBN: 9780170389433

2 From each branch end, create new branches for routes that are **likely** to get you nearer to the finish. You can ignore routes that are obviously longer.

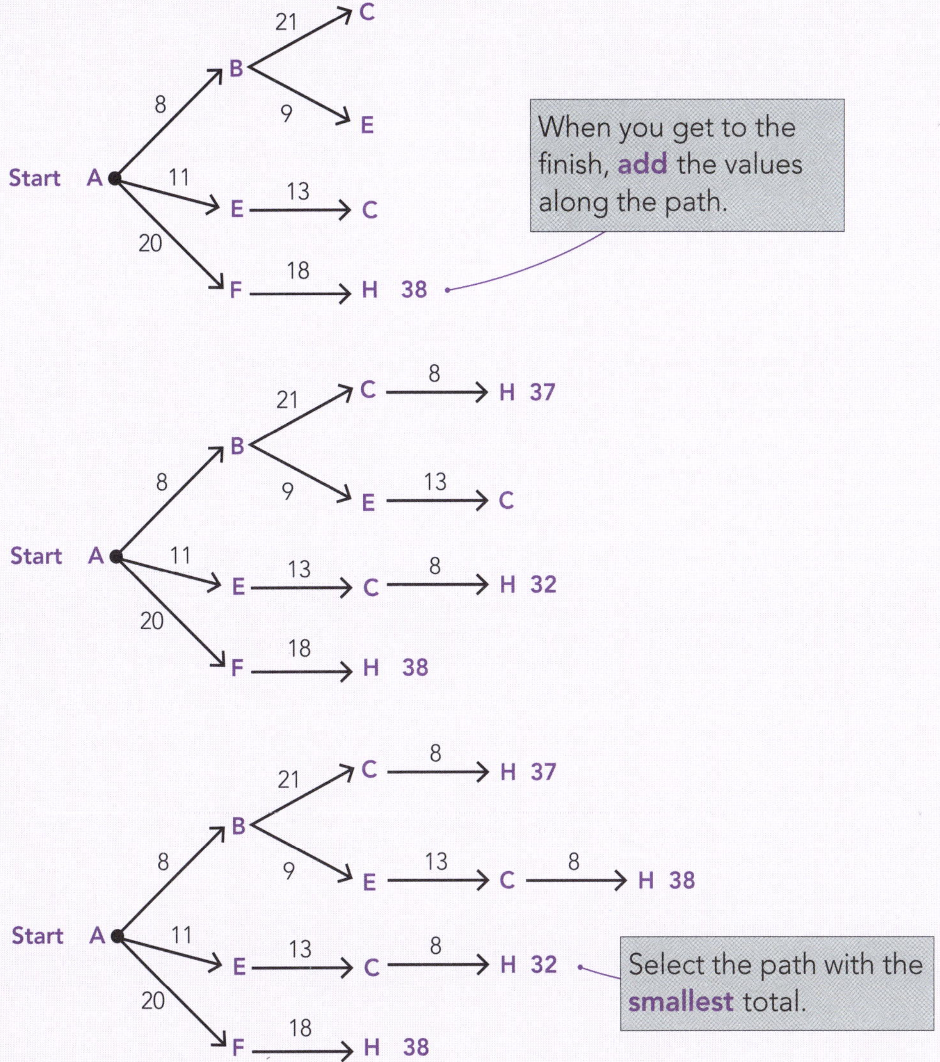

Shortest route: A → E → C → H

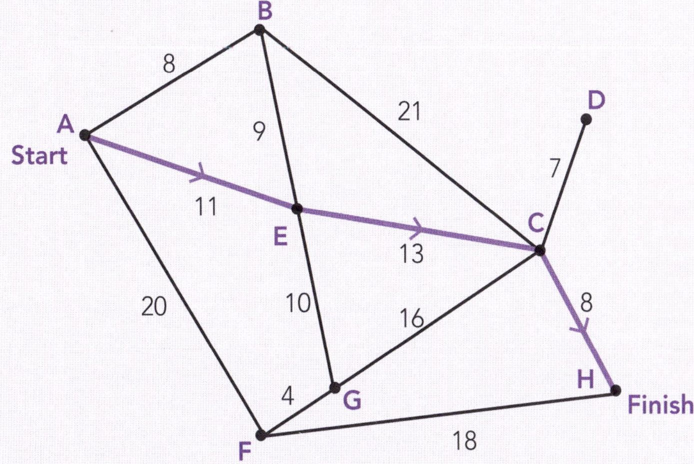

Weighted value of the shortest route = 32 m

ISBN: 9780170389433

Highlight the shortest paths on the diagrams and give their weighted values.

1 The diagonal distance between adjacent dots is 1 unit.

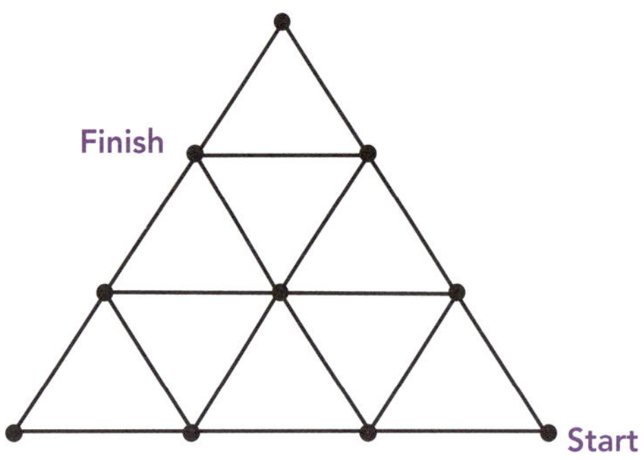

Shortest path length: ______________

2 The diagram shows distances between towns.

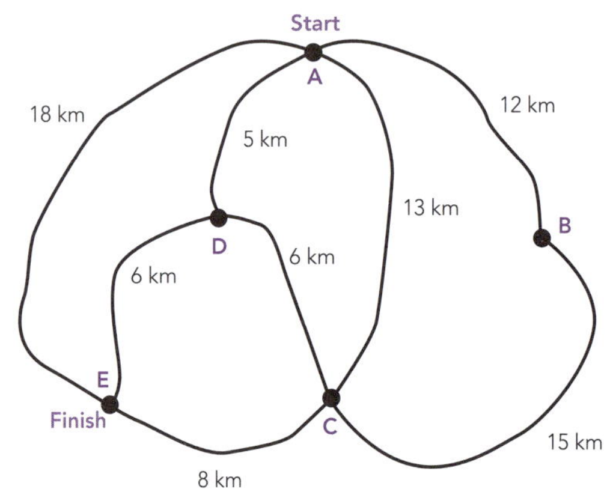

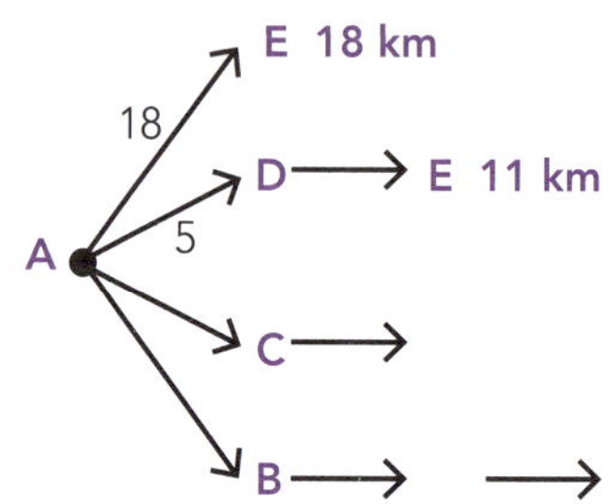

Shortest path length: ______________

3 Times (hours) taken to tramp each track in a network within a national park.

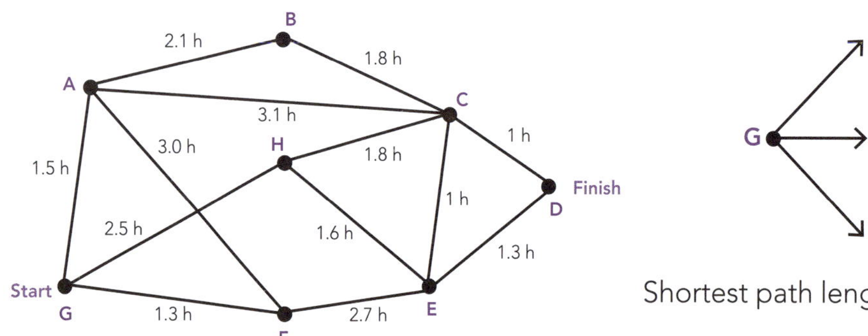

Shortest path length: ______________

4 The diagram shows the times (minutes) taken to walk a series of tracks in a park.

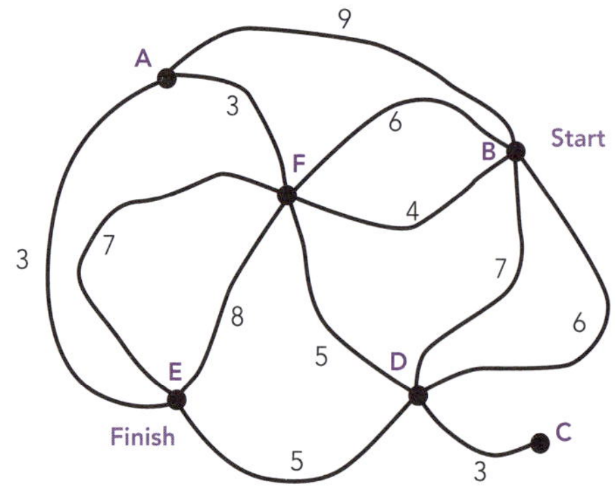

Shortest path length: ______________

ISBN: 9780170389433

5 The distance between each node is given in metres. List the shortest paths and find their lengths.

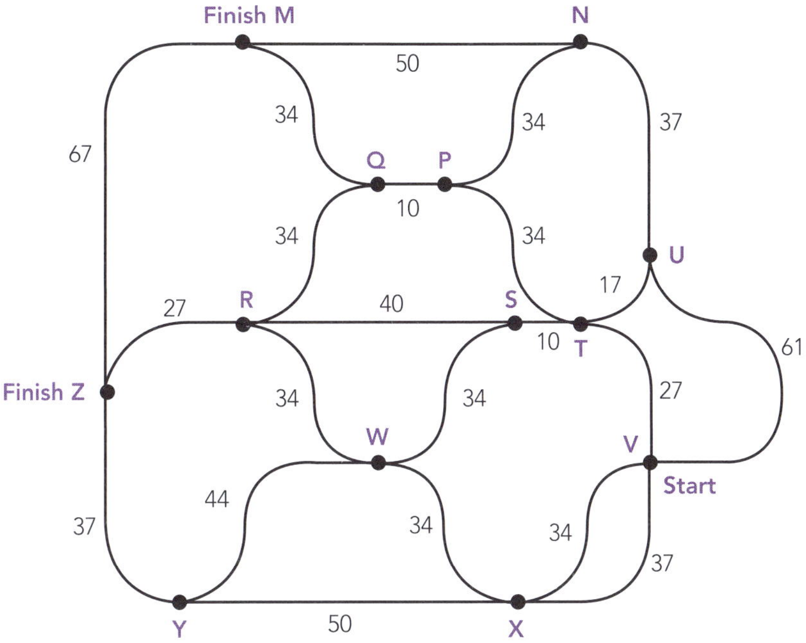

a Shortest path from the start (V) to finish (Z): ______________________

Length of shortest path from the start (V) to finish (Z): ____________

b Shortest path from the start (V) to finish (M): ______________________

Length of shortest path from the start (V) to finish (M): ____________

ISBN: 9780170389433

6 This network shows the distances by road between some South Island towns. Describe the shortest routes from Greymouth to Murchison, and Greymouth to Picton, and give their lengths.

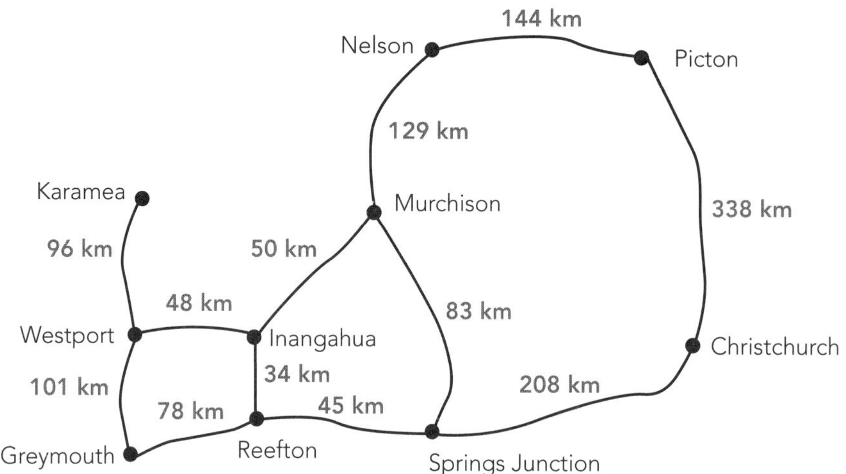

a Greymouth to Murchison description: ______________________________

Shortest path from Greymouth to Murchison: ______________________

b Greymouth to Picton description: ______________________________

____

Shortest path from Greymouth to Picton: ______________________

 ISBN: 9780170389433

7 The map shows the distances (km) between cities. Describe the shortest routes between the locations below, and find their lengths.

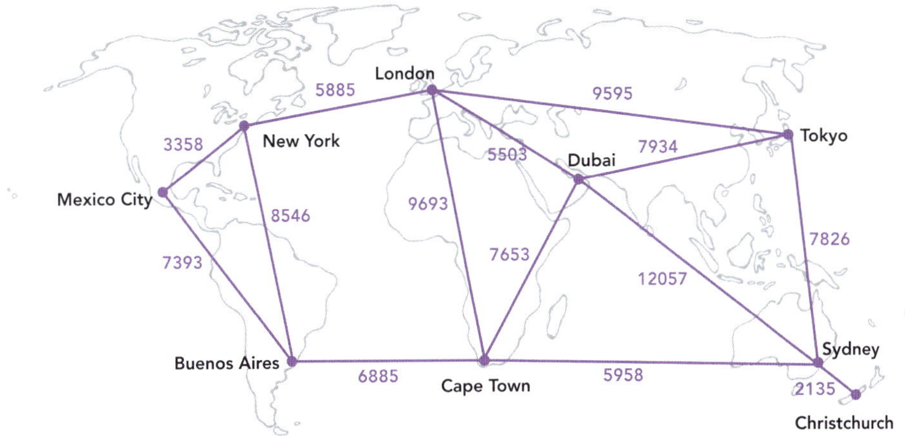

a

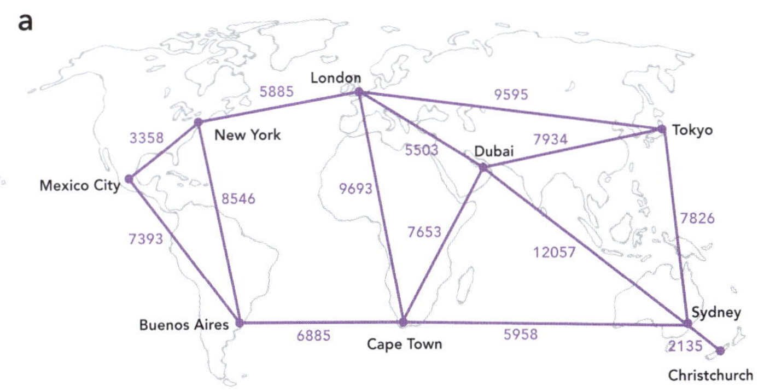

Shortest route from London to Sydney: ____________________

____

Length: ____________________

b

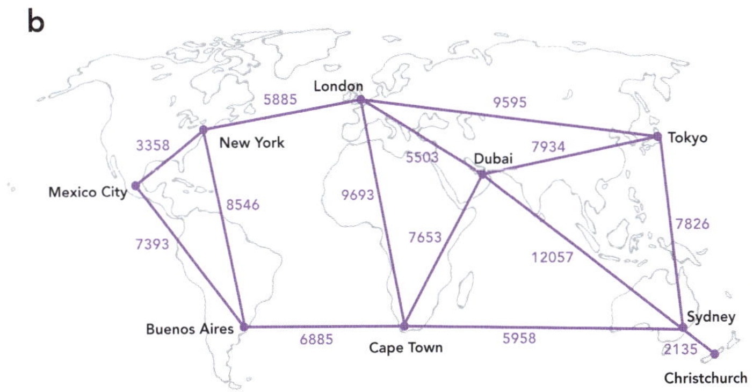

Shortest route from Cape Town to New York: ____________________

____

Length: ____________________

ISBN: 9780170389433

c

Shortest route from Cape Town to Tokyo: ______________________________

____

Length: ______________________

d

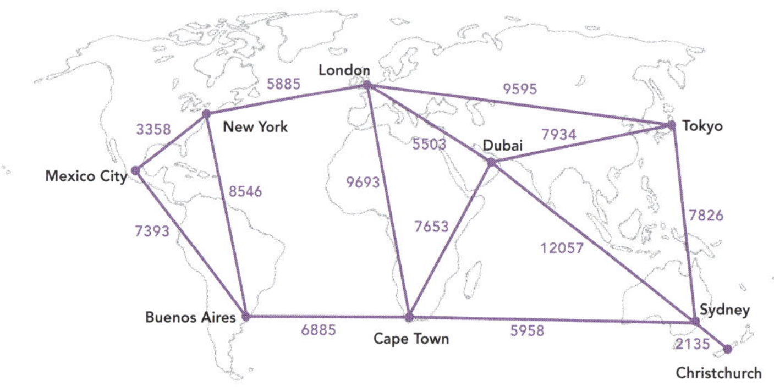

Shortest route from Bueno Aires to Tokyo: ______________________________

____

Length: ______________________

e

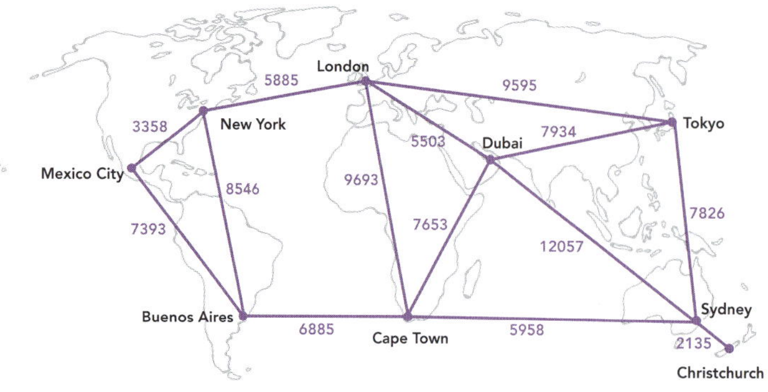

Shortest route from Mexico City to Sydney: ______________________________

____

Length: ______________________

ISBN: 9780170389433

Traversability

Rules:

- A network is traversable if you can start at one node and trace **every edge exactly once**.
- You may visit a **node more than once**.
- You may or may not end up where you started.
- A network is either traversable or not.

A network is traversable if it has either **no odd nodes** or exactly **two odd nodes**.

There are two types of traversable networks:

1 **Euler circuits**
- Have **no odd nodes**.
- End up where they started.
- It doesn't matter where you start.

Note: 'Euler' is pronouncd 'oiler'.

Example:

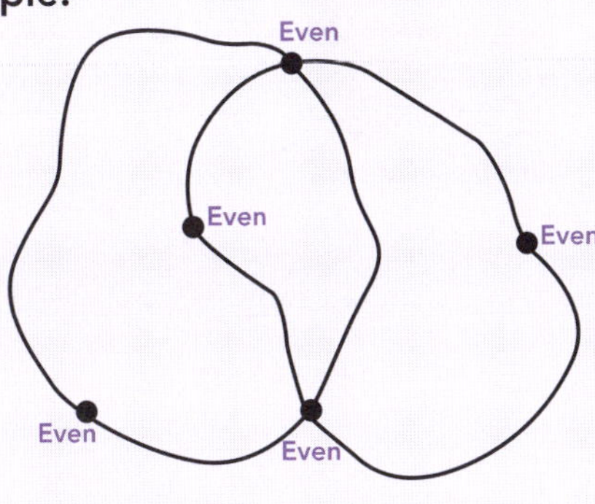

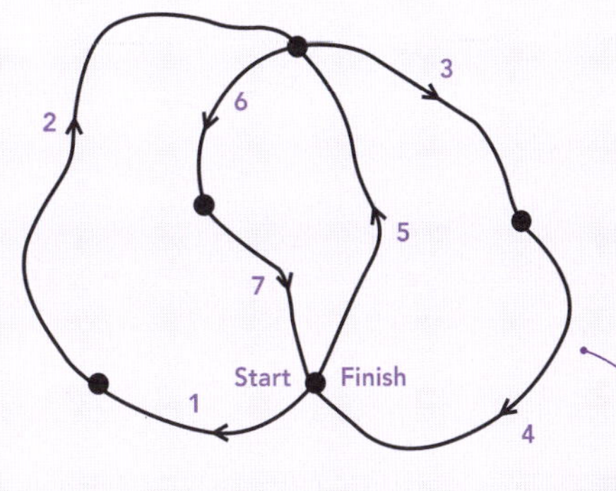

No odd nodes ⇒ Euler circuit

These are examples: there are several ways of doing both of these.

2 **Euler paths**
- Have **exactly two odd nodes**.
- Do not end up where they started.
- Start and finish at odd nodes.

Example:

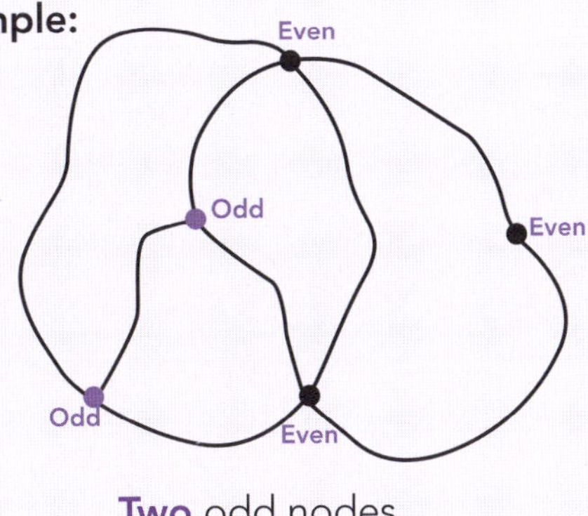

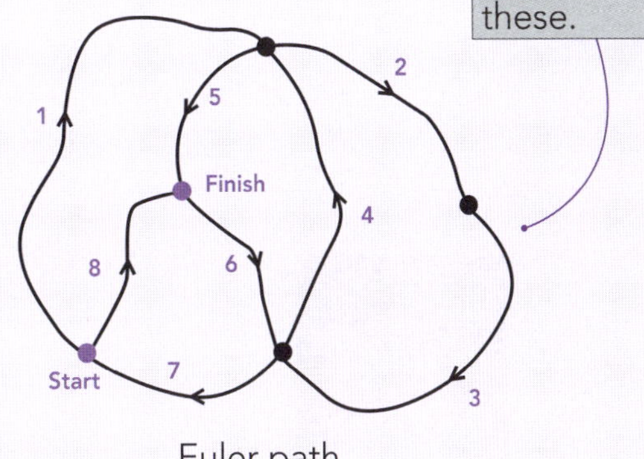

Two odd nodes ⇒ Euler path

ISBN: 9780170389433

For the following networks: Identify the odd nodes.
State whether the network is traversable or not and why.
If the network is traversable, state whether it is an Euler path or an Euler circuit, and show one possible route on the diagram.

1

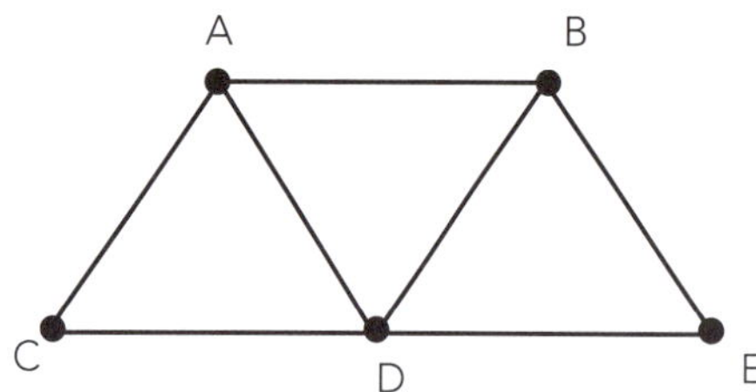

Node	A	B	C	D	E
Degree	3				
Odd/Even	Odd				

Traversable? Yes/No

Reason: ____

If Yes: Euler path/Euler circuit

2

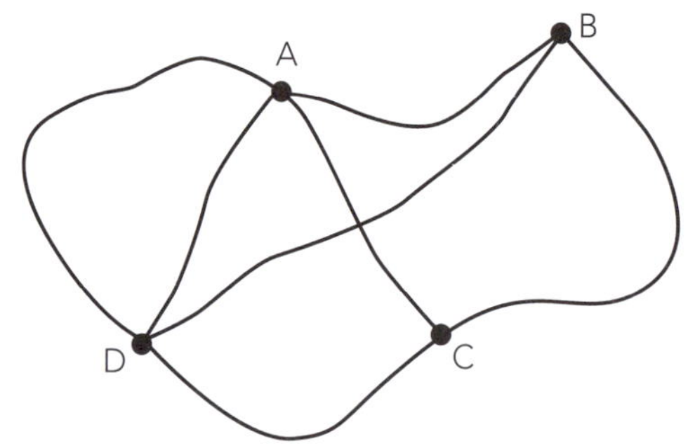

Node	A	B	C	D
Degree				
Odd/Even				

Traversable? Yes/No

Reason: ____

If Yes: Euler path/Euler circuit

3

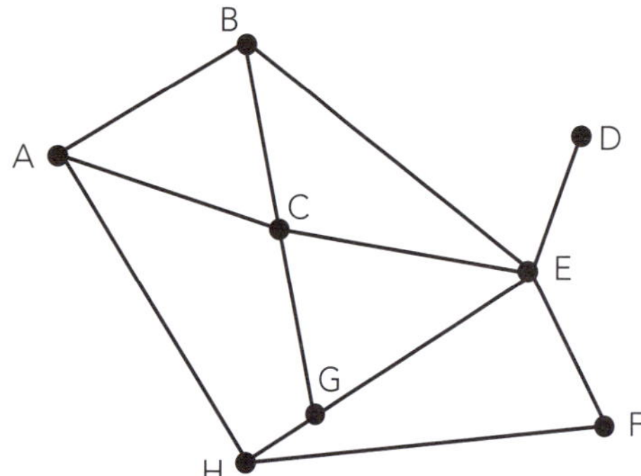

Traversable? Yes/No

Reason: ____

____

If Yes: Euler path/Euler circuit

Node								
Degree								
Odd/Even								

ISBN: 9780170389433

4

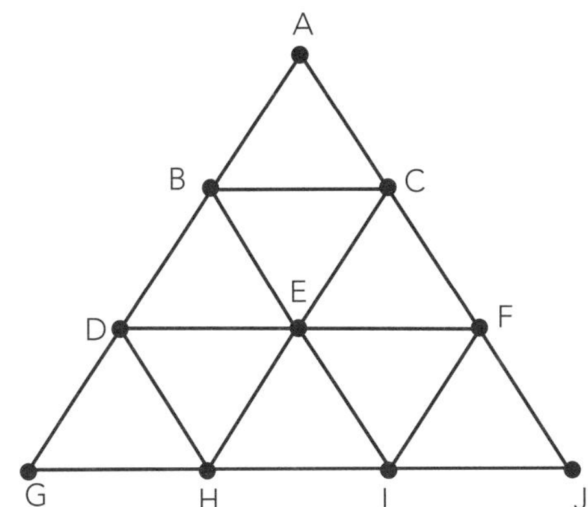

Traversable? Yes/No

Reason: ____________________

____

____

If Yes: Euler path/Euler circuit

5

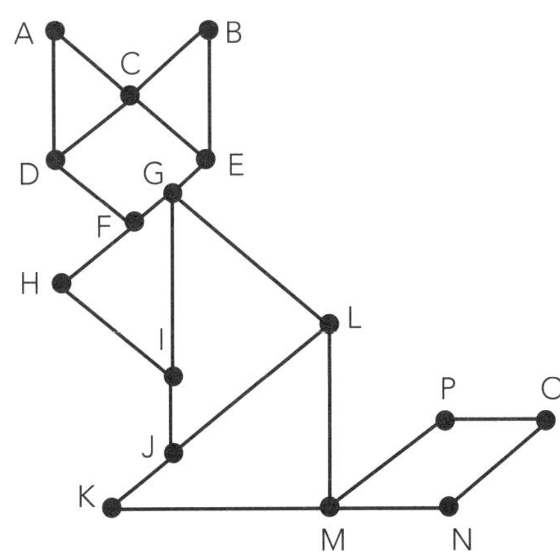

Traversable? Yes/No

Reason: ____________________

____

____

If Yes: Euler path/Euler circuit

6

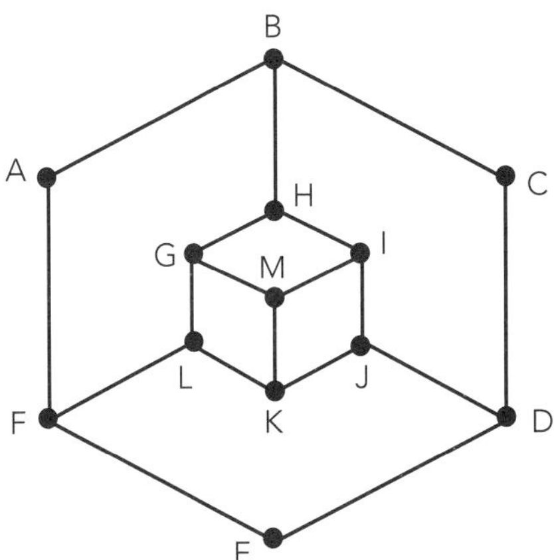

Traversable? Yes/No

Reason: ____________________

____

____

If Yes: Euler path/Euler circuit

ISBN: 9780170389433

7

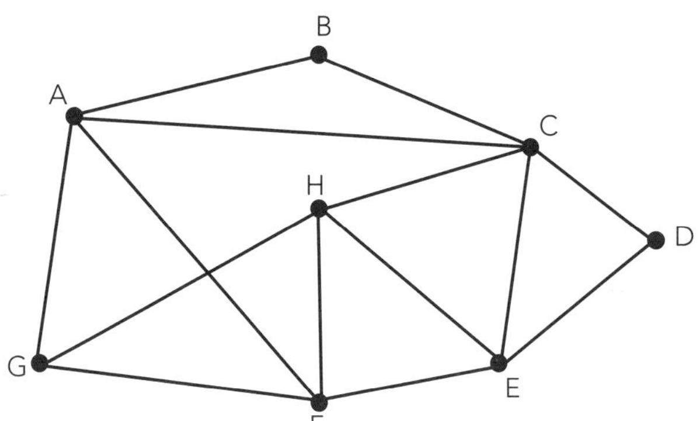

Traversable? Yes/No

Reason: ______________________________

____

____

If Yes: Euler path/Euler circuit

8

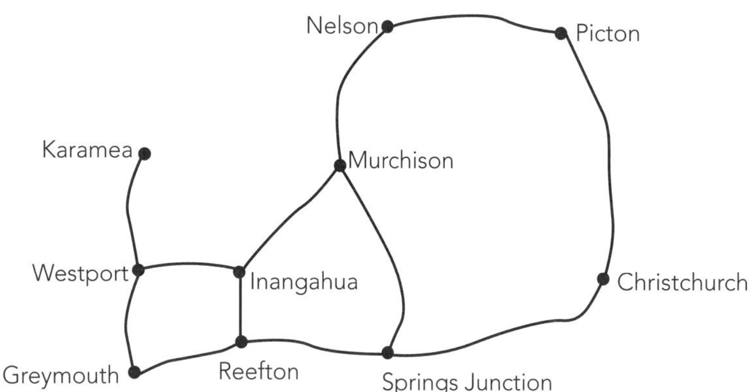

Traversable? Yes/No

Reason: ______________________________

____

____

If Yes: Euler path/Euler circuit

9

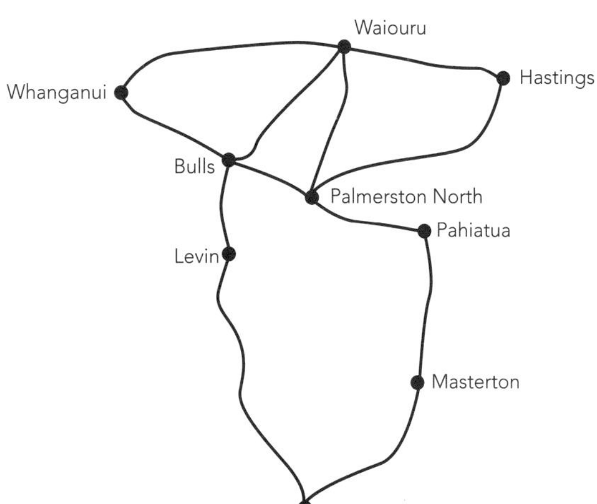

Traversable? Yes/No

Reason: ______________________________

____

____

If Yes: Euler path/Euler circuit

ISBN: 9780170389433

10 For each Platonic solid, state whether it is traversable, and give a reason for each answer. Note: all the faces on Platonic solids are identical, so all the nodes and edges are identical.

a

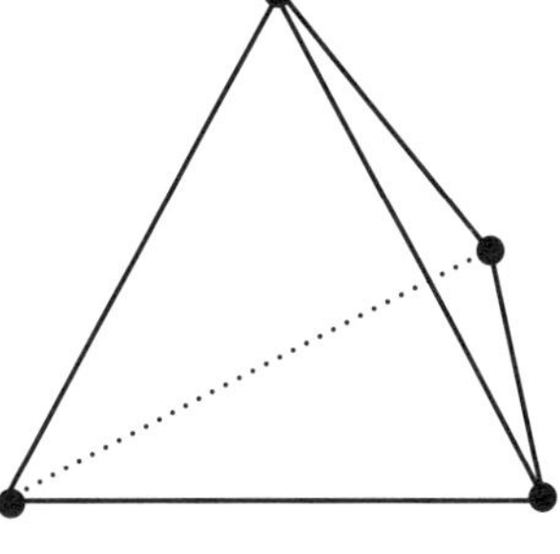

Traversable? Yes/No

Reason: ______________________________

____

b

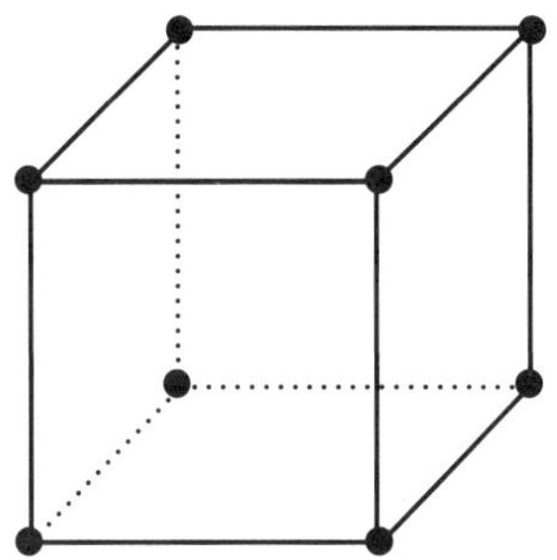

Traversable? Yes/No

Reason: ______________________________

____

c

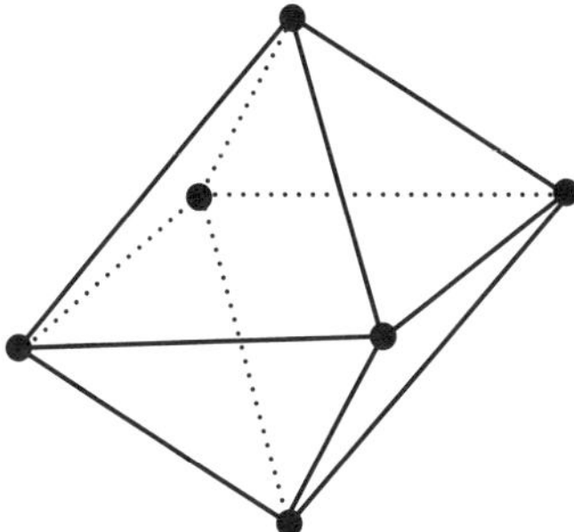

Traversable? Yes/No

Reason: ______________________________

____

d

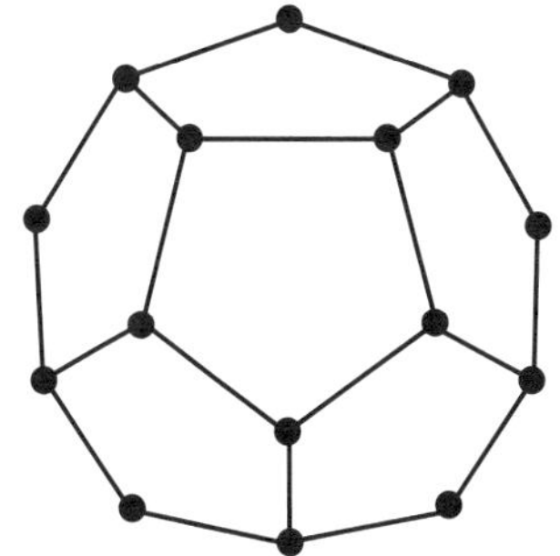

Traversable? Yes/No

Reason: ______________________________

____

e

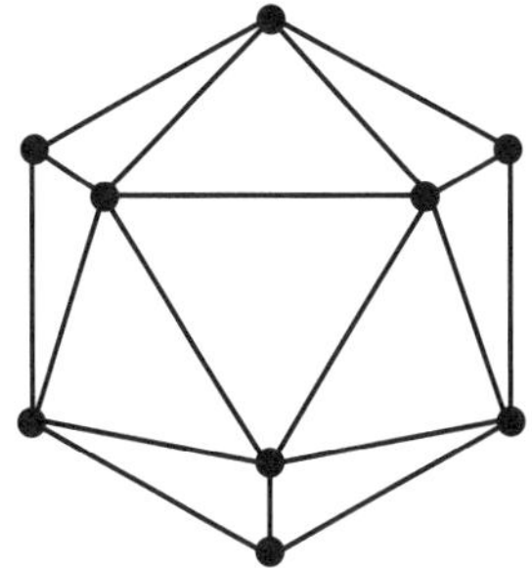

Traversable? Yes/No

Reason: ______________________________

____

Putting it all together

Answer the following.

1 This network shows the locations of Auckland and some Pacific islands, with the times needed to fly between locations.

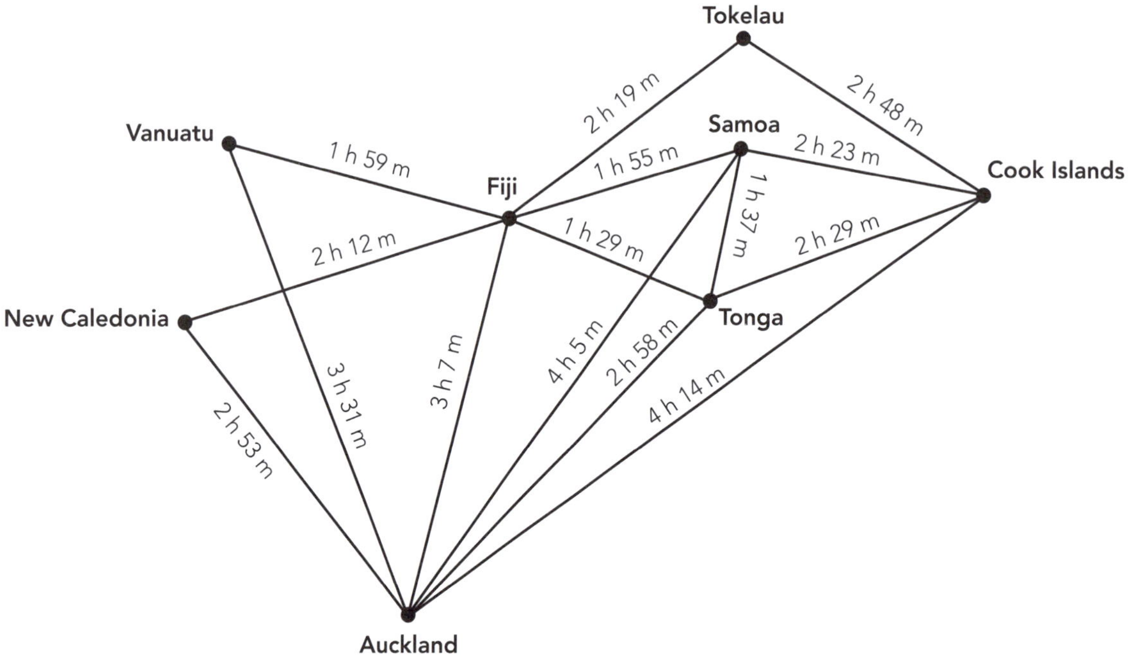

a Find the shortest route between New Caledonia and the Cook Islands. How much flying time is involved in this trip?

Warning: Be careful when adding times. 1 h 59 m ≠ 1.59 hours!

Route: ____

Total flying time: ________________________

ISBN: 9780170389433

b This network of flights is run by a company based in Fiji. The new chief executive would like to fly every route exactly once in order to discover what customers are experiencing. He would like to start and finish his trip in Fiji. Is this possible? If so, suggest a route he could take. You may like to complete the table to support your discussion.

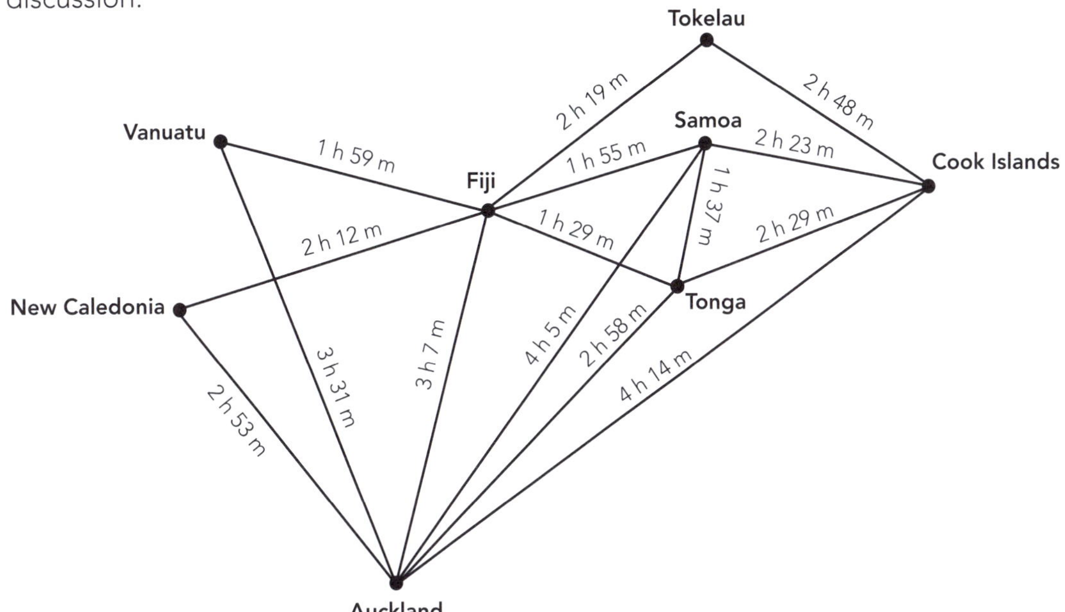

Node	Degree	Odd/Even
Vanuatu		
Auckland		
New Caledonia		
Fiji		
Tonga		
Tokelau		
Cook Islands		
Samoa		

ISBN: 9780170389433

2 This network shows the paths connecting blocks of buildings at a school. The times (in minutes) needed to walk between each block are written on each path.

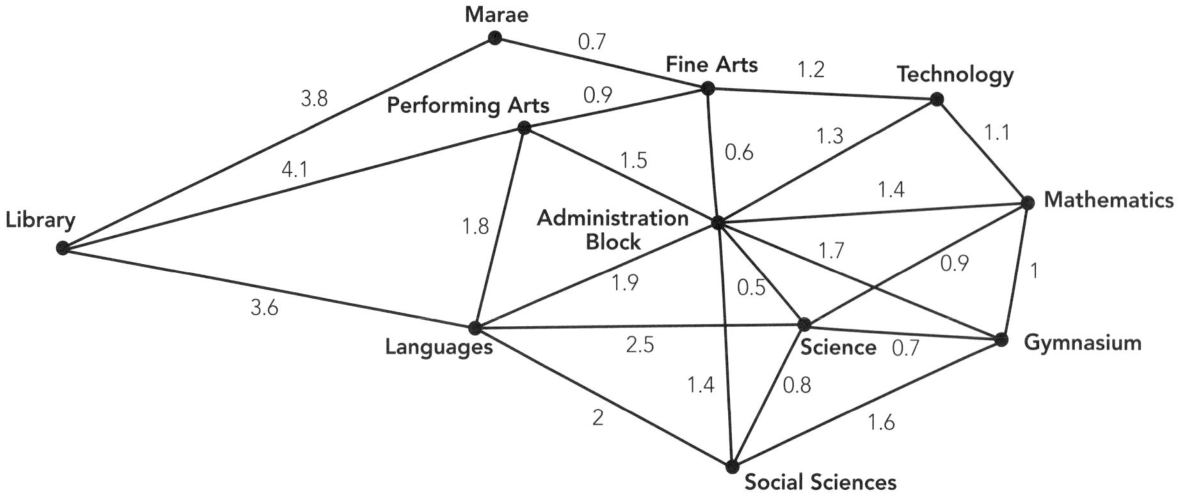

a Find the shortest route(s) from the Gymnasium to the Performing Arts block.

b Mrs Walker needs to get from the Mathematics block to the Library within 5 minutes, or she will be late for her class. Can she make it in time? If not, how late will she be if she takes the shortest route possible?

 ISBN: 9780170389433

c The principal likes to do a weekly walkabout of the buildings. Ideally he would like to start from his office in the Administration Block and walk every path exactly once, and return to his office. He has been told that this is not possible. Explain why he has been told this.

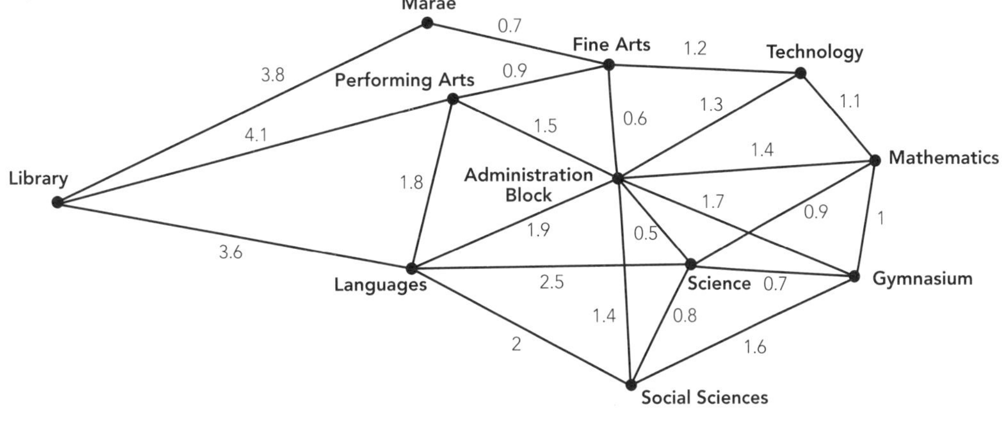

d He would like to add one path to the network in order to make it possible for him to walk every path exactly once, but not necessarily starting and finishing at the Administration Block. Add a possible path to the network, and explain your reasoning.

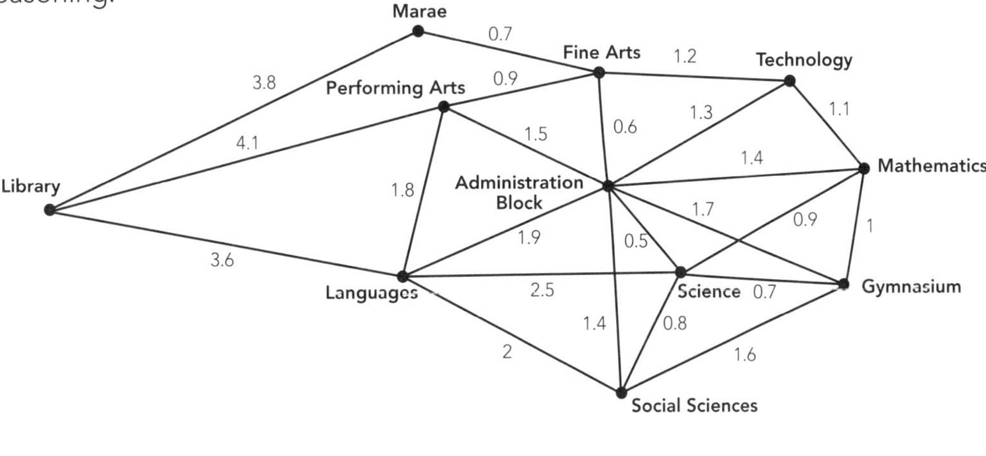

ISBN: 9780170389433

Networks from tables

- The values of edges in a network are often called **'weightings'**.
- Weightings for a network can be given in table form.

Examples of weightings:
- distances on an orienteering map
- costs of construction of each section of a track network in a national park
- the time it takes to drive between each city on a map.

Example 1: The chart shows the distances (km) between points.

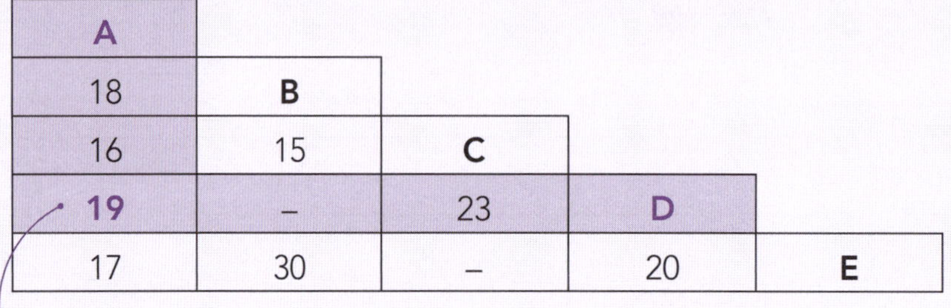

A				
18	B			
16	15	C		
19	–	23	D	
17	30	–	20	E

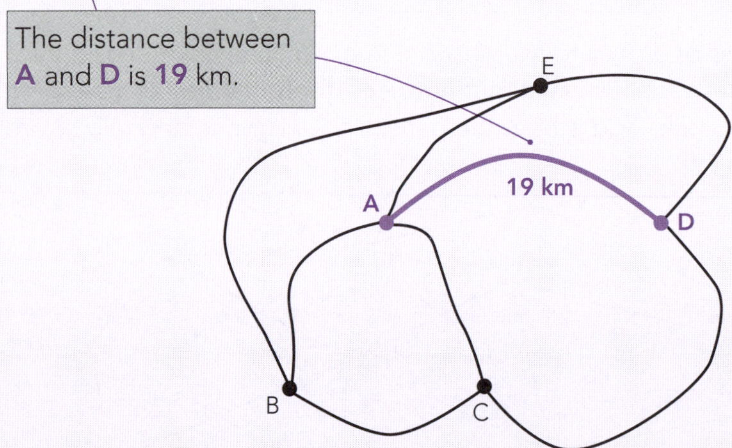

The completed map:

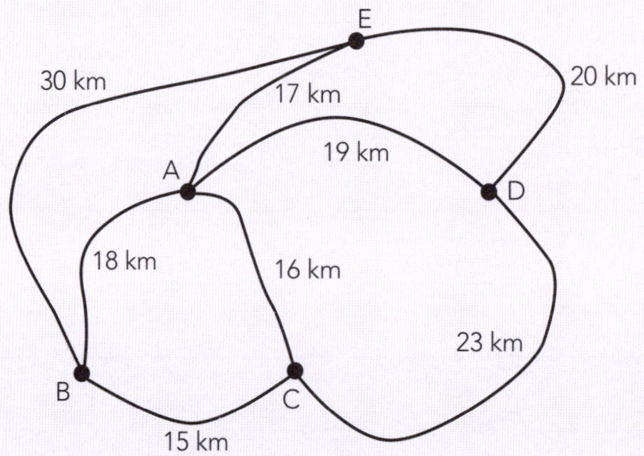

ISBN: 9780170389433

Example 2: This is a different type of chart. It shows the distances (km) between some towns.

Notice that every distance appears twice, and that each side of the diagonal is a **mirror image** of the other.

	Karamea	Westport	Greymouth	Inangahua	Reefton	Murchison	Springs Junction	Nelson
Karamea		96						
Westport	96		101	**48**				
Greymouth		101			78			
Inangahua		**48**			34	50		
Reefton			78	34			45	
Murchison				50			83	129
Springs Junction					**45**	83		
Nelson						129		

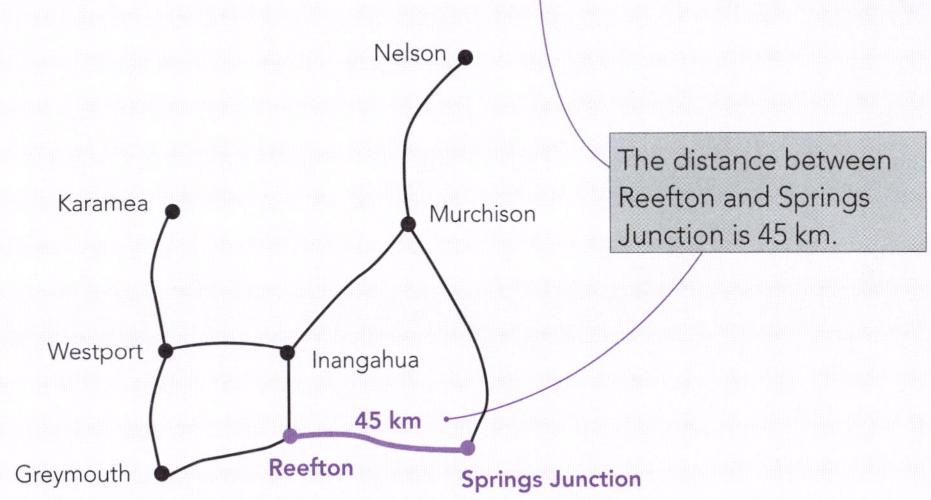

The distance between Reefton and Springs Junction is 45 km.

The completed map:

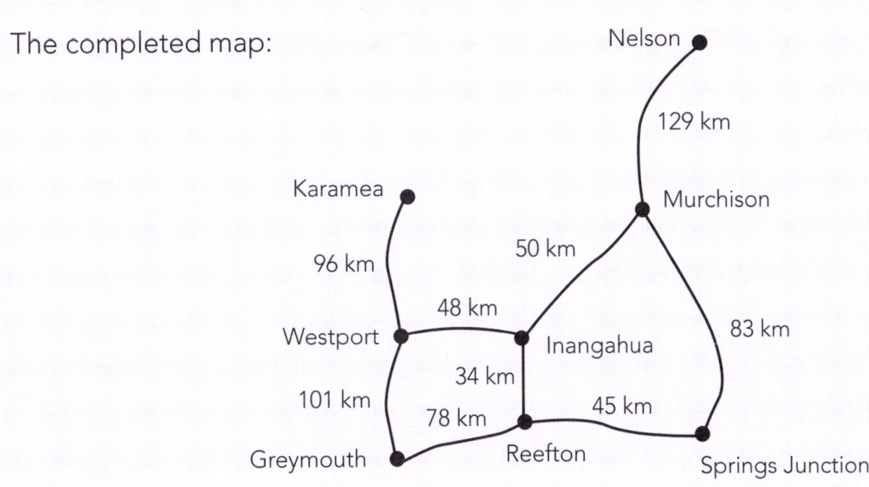

ISBN: 9780170389433

Example 3: This chart shows the costs (in $1000s) of connecting some locations with fibre.

	A	B	C	D	E	F	G
A		18		19			
B	18		12		9	22	17
C		12			15		10
D	19				20		26
E		9	15	20			
F		22					
G		17	10	26			

Draw a network to show this.

Step 1: Select the node that connects to the largest number of other nodes, and place this at the centre of your network.

	A	B	C	D	E	F	G
A		18		19			
B	18		12		9	22	17
C		12			15		10
D	19				20		26
E		9	15	20			
F		22					
G		17	10	26			

B has connections to five nodes.

∴ Place **B** at the centre of your network and arrange the five nodes it connects to around it.

Step 2: Place any remaining nodes to one side, and then use the table to add edges and their lengths.

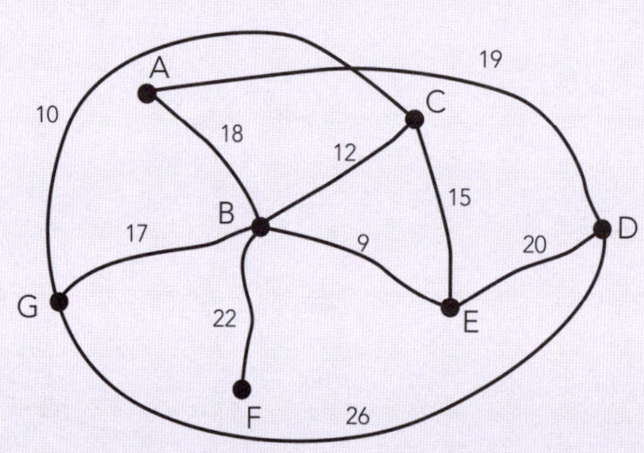

Note:

1 You do not need to make the edges the correct lengths, or draw your diagram to scale.

2 Your diagram may look rather messy by the time you have added all the nodes and edges.

ISBN: 9780170389433

Add the missing distances to the following tables and networks.

1

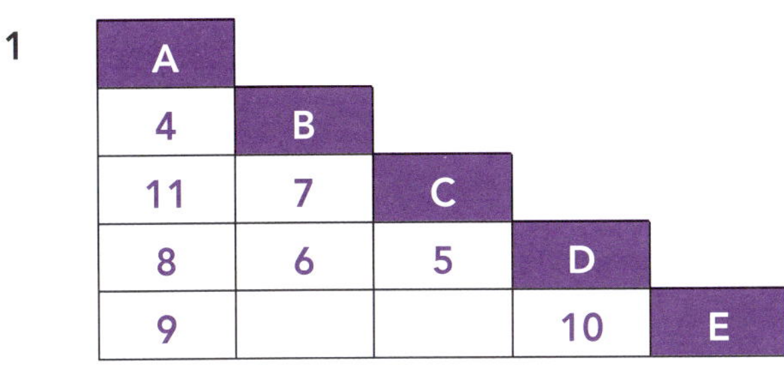

A				
4	B			
11	7	C		
8	6	5	D	
9			10	E

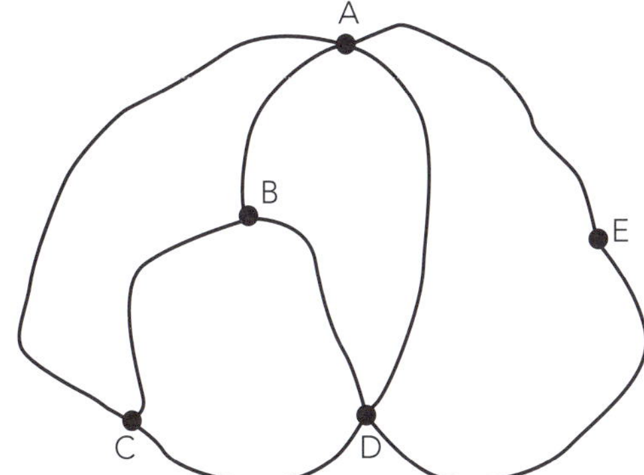

2

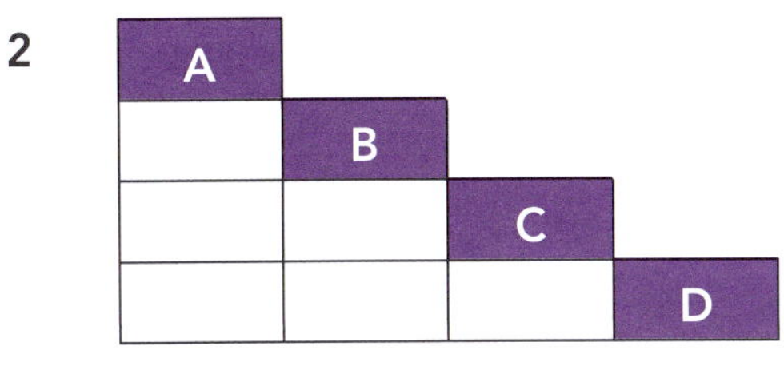

A			
	B		
		C	
			D

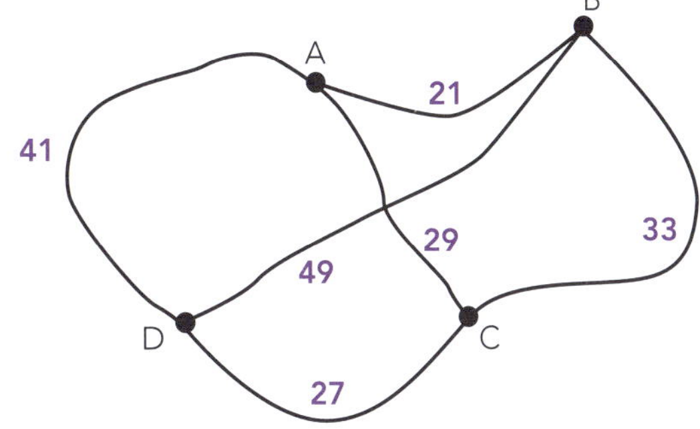

3

	A	B	C	D	E	F	G
A			51		38	33	
B			22	24		40	
C	51	22		26			
D		24	26		18		25
E	38			18			11
F	33	40					42
G				25	11	42	

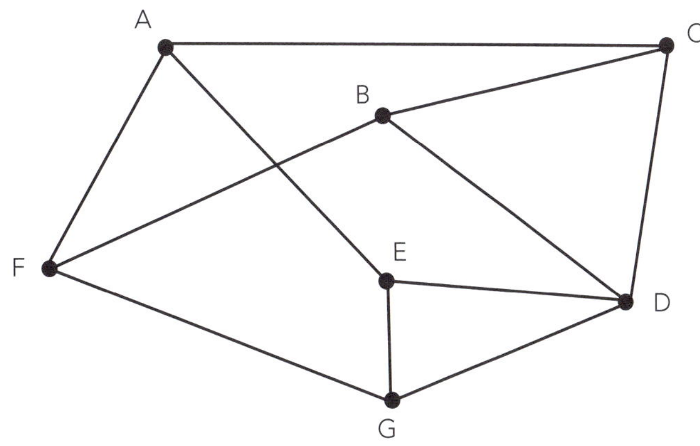

4

	A	B	C	D	E	F	G	H
A								
B								
C								
D								
E								
F								
G								
H								

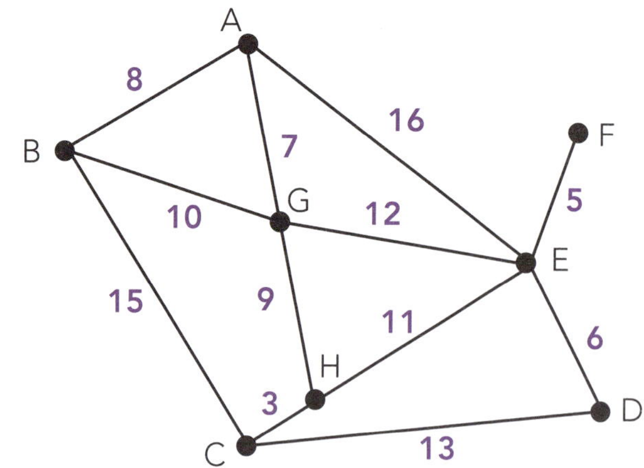

5

	Wellington	Levin	Masterton	Woodville	Palmerston North	Bulls	Whanganui	Waiouru
Wellington		93	99					
Levin	93				48	56		
Masterton	99			80				
Woodville			80		55			
Palmerston North		48		55		29		
Bulls		56			29		44	110
Whanganui						44		107
Waiouru						110	107	

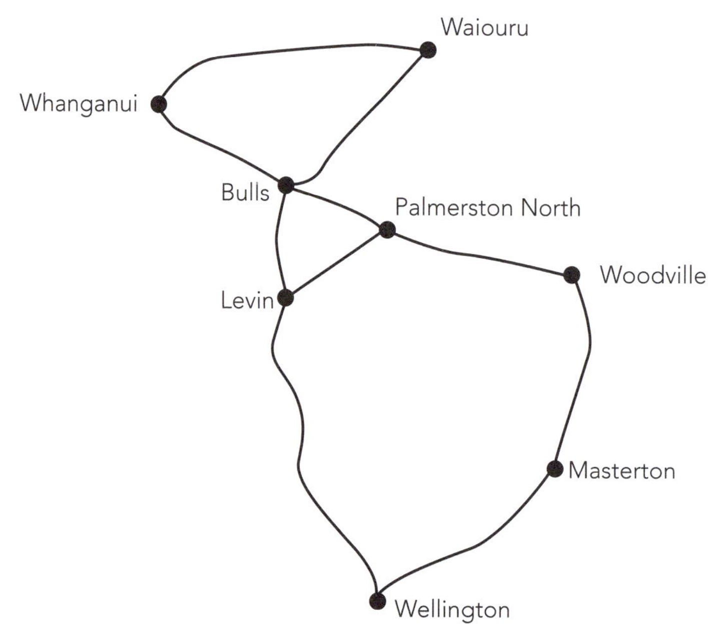

ISBN: 9780170389433

6 The map shows the distances between cities (km).

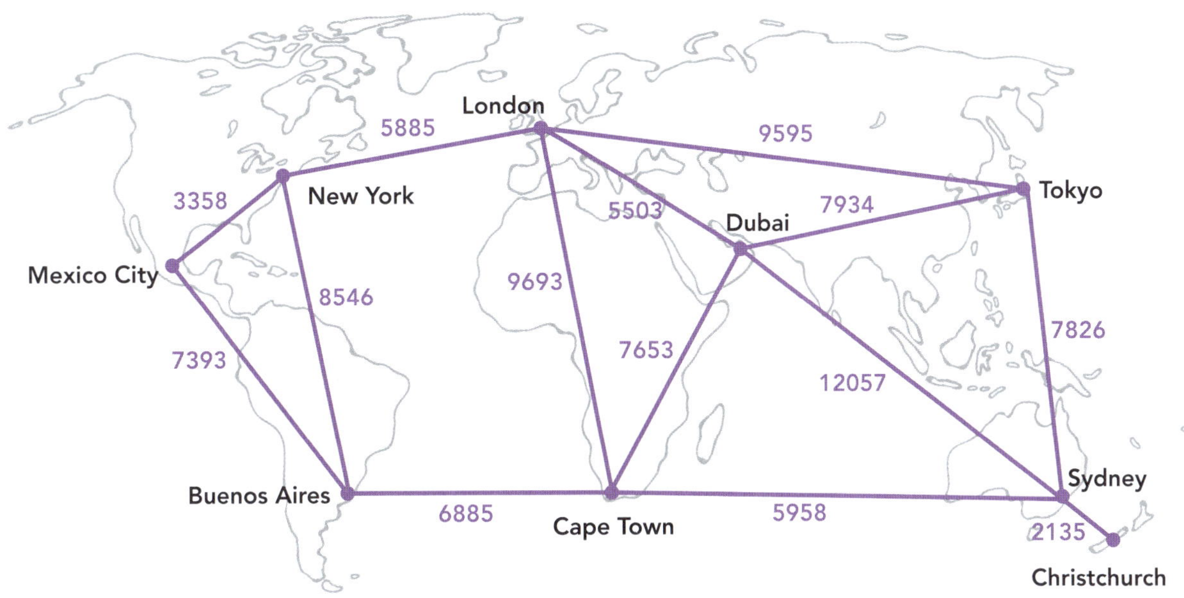

a Complete the table to show the distances between cities.

	Mexico City	Buenos Aires	New York	London	Cape Town	Dubai	Tokyo	Sydney	Christchurch
Mexico City									
Buenos Aires									
New York									
London									
Cape Town									
Dubai									
Tokyo									
Sydney									
Christchurch									

b Add to the map the time taken (hours and minutes) to fly between cities.

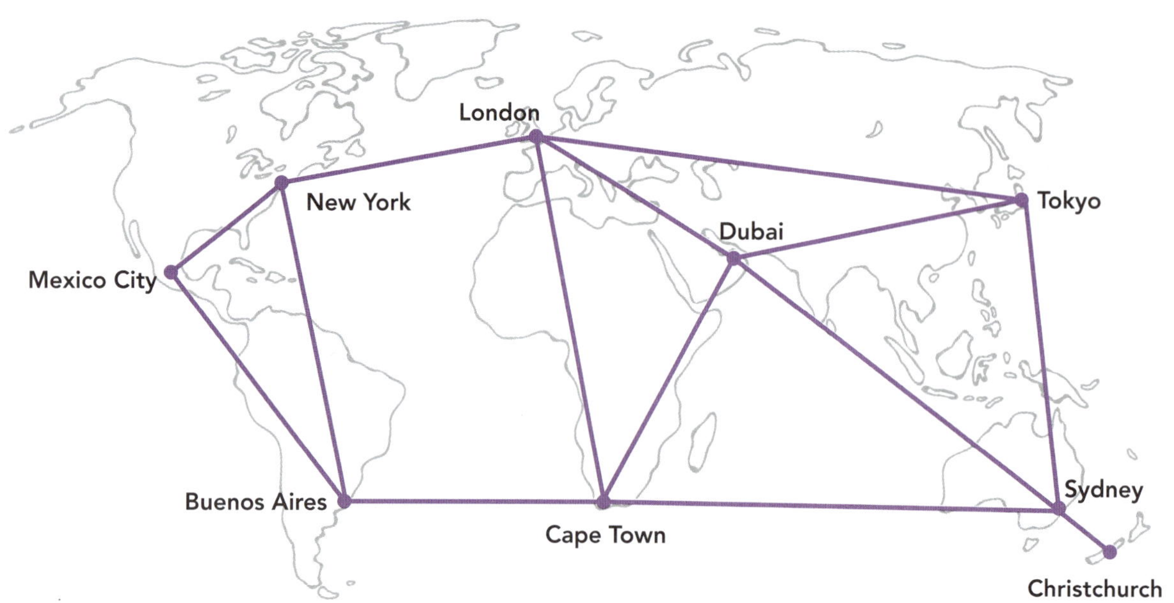

	Mexico City	Buenos Aires	New York	London	Cape Town	Dubai	Tokyo	Sydney	Christchurch
Mexico City		9 h 55 m	4 h 45 m						
Buenos Aires	9 h 55 m		10 h 55 m		18 h 55 m				
New York	4 h 45 m	10 h 55 m		8 h 15 m					
London			8 h 15 m		11 h 30 m	6 h 55 m	12 h 45 m		
Cape Town		18 h 55 m		11 h 30 m		9 h 40 m		17 h 45 m	
Dubai				6 h 55 m	9 h 40 m		11 h 50 m	13 h 45 m	
Tokyo				12 h 45 m		11 h 50 m		9 h 35 m	
Sydney					17 h 45 m	13 h 45 m	9 h 35 m		3 h 5 m
Christchurch								3 h 5 m	

ISBN: 9780170389433

Spanning trees

- A spanning tree forms connections to **every node**.
- It does **not** have to be a path or circuit.
- It does **not** need to use every edge.

Minimum spanning trees

- A minimum spanning tree connects all the nodes using the **minimum** weighting of edges possible.
- There can be **more than one** minimum spanning tree.
- Examples of uses:
  - to connect electrical sockets in a house with a minimum length or cheapest cost of wiring
  - to connect a series of towns with the shortest total length of road.

Example:

Step 1: Make an **ordered** list of all the path lengths.

15	16	17	18	19	20	22	23	30

Step 2: Highlight any paths that lead to nodes with the value of 1. You **must** use these.

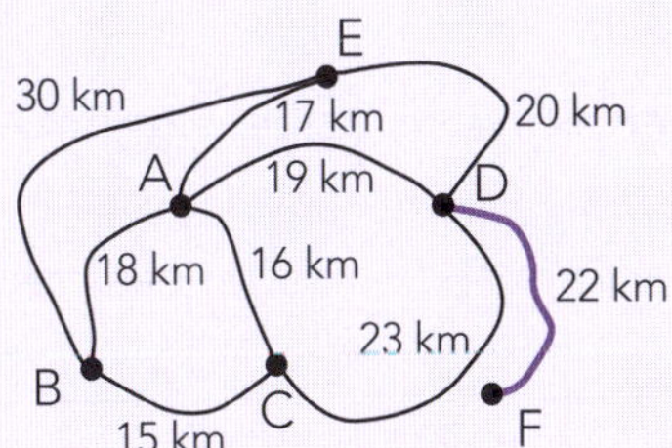

15	16	17	18	19	20	~~22~~	23	30

Step 3: Highlight the shortest paths on your network, crossing each one off as you go.

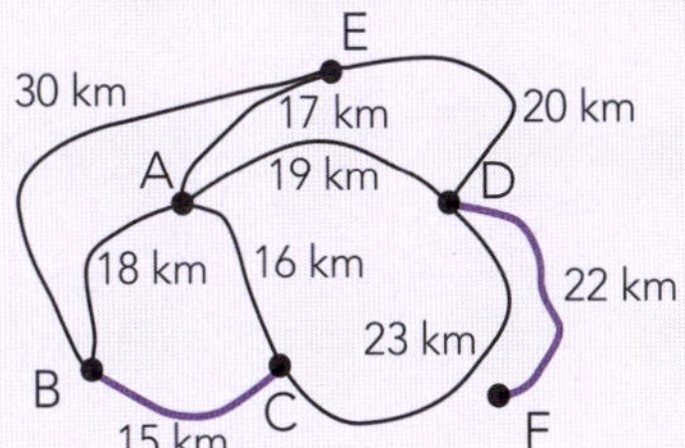

~~15~~	16	17	18	19	20	~~22~~	23	30

ISBN: 9780170389433

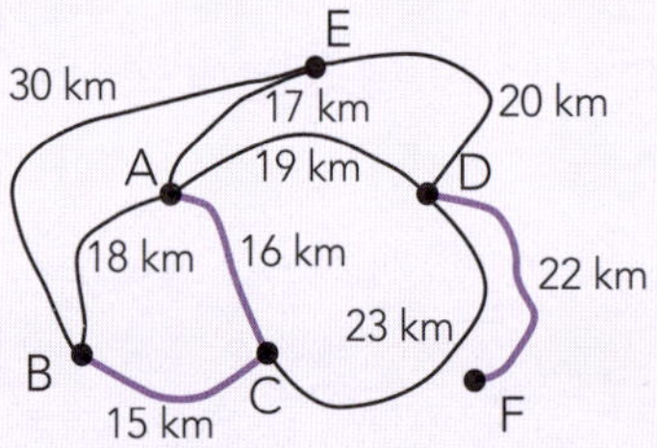

~~15~~	~~16~~	17	18	19	20	~~22~~	23	30

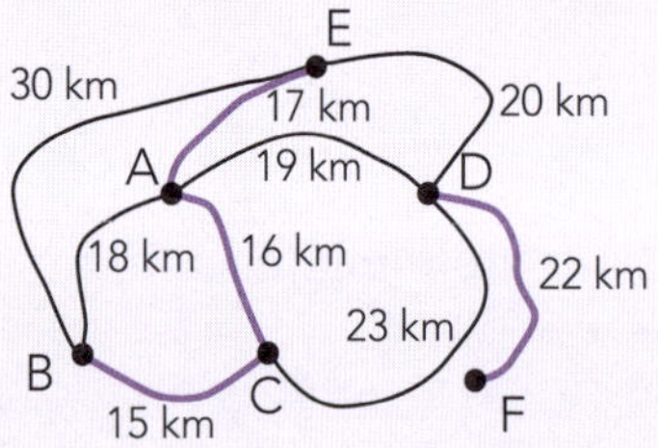

~~15~~	~~16~~	~~17~~	18	19	20	~~22~~	23	30

There is no point in using the 18 because A and B are already connected via C.

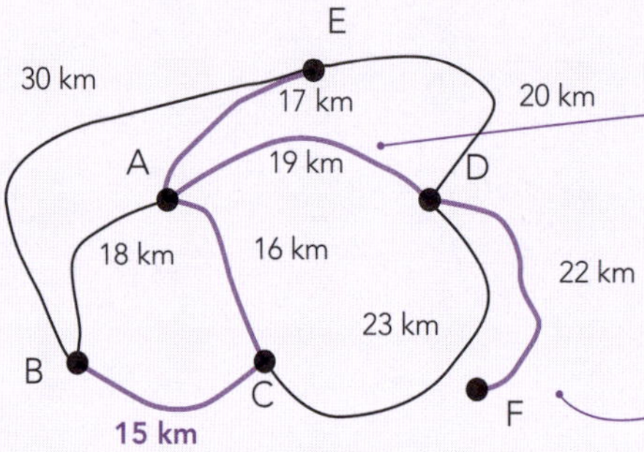

This is the shortest connection from D to the rest of the spanning tree.

Finished minimum spanning tree: every node can be reached by following the edges that form the spanning tree.

Step 4: Calculate the total length of the roads needed to connect all the towns:
Length = 15 + 16 + 17 + 19 + 22 = **89 km**.

Highlight the minimum spanning tree for each diagram, and give its value, with units.

1

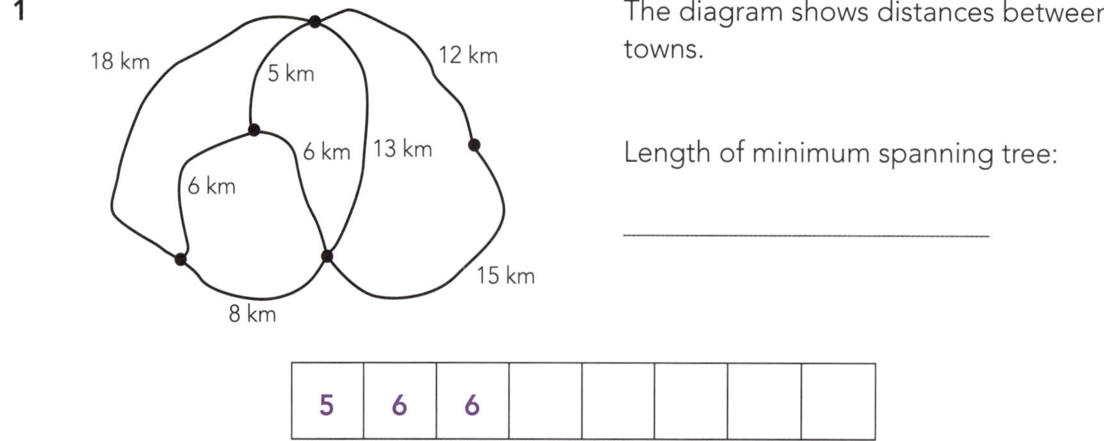

The diagram shows distances between towns.

Length of minimum spanning tree:

____

5	6	6					

 ISBN: 9780170389433

2 The diagram shows the times (minutes) taken to walk a series of tracks in a park.

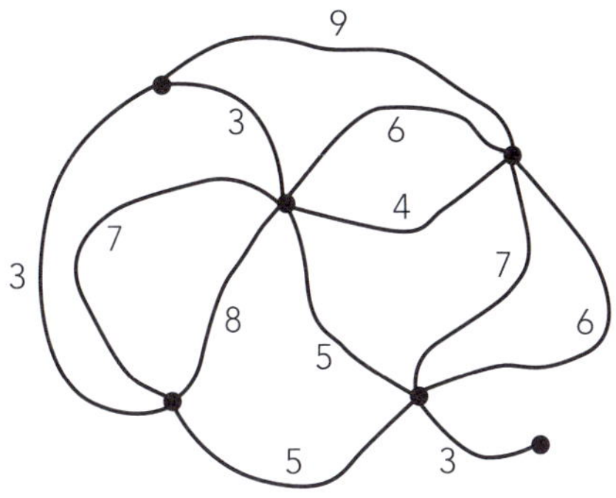

Length of minimum spanning tree:

____

3

3 The diagram shows the distances (nautical miles) between anchorages in a marine park.

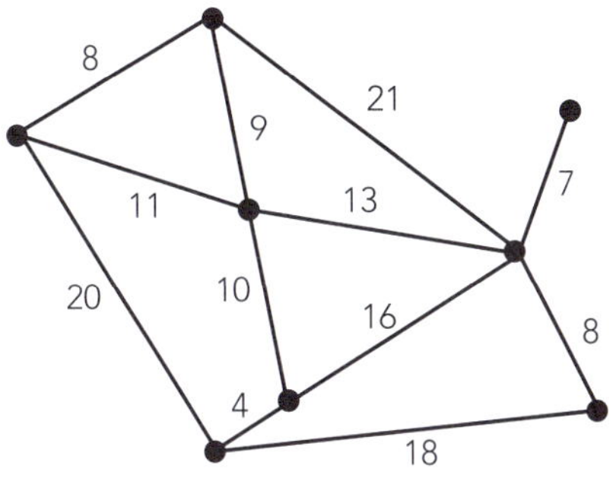

Length of minimum spanning tree:

____

4 The diagram shows the distances between some South Island towns.

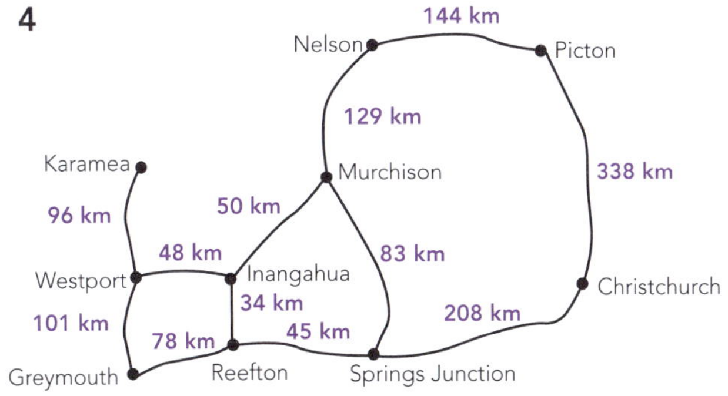

Length of minimum spanning tree:

____

5 The diagram shows the time needed each week to clear sections of tracks in a national park.

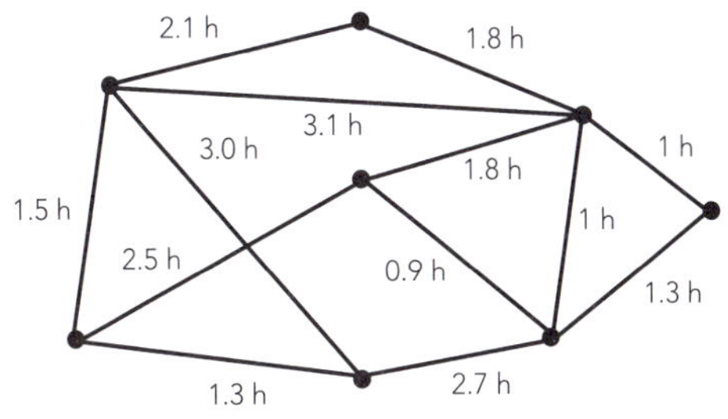

Length of minimum spanning tree:

____

ISBN: 9780170389433

6

107 km
Waiouru
Whanganui
110 km
44 km
Bulls
29 km
Palmerston North
56 km
Levin
48 km
55 km
Woodville
80 km
93 km
Masterton
99 km
Wellington

The diagram shows the distances between some North Island towns.

Length of minimum spanning tree: ______________________

7 The map shows the distances between cities (km). Leila dislikes flying, so she would like to visit every location on the map by travelling the shortest distance possible.

a On the map, highlight the shortest flights needed to connect every location.

b Calculate its length. ______________________

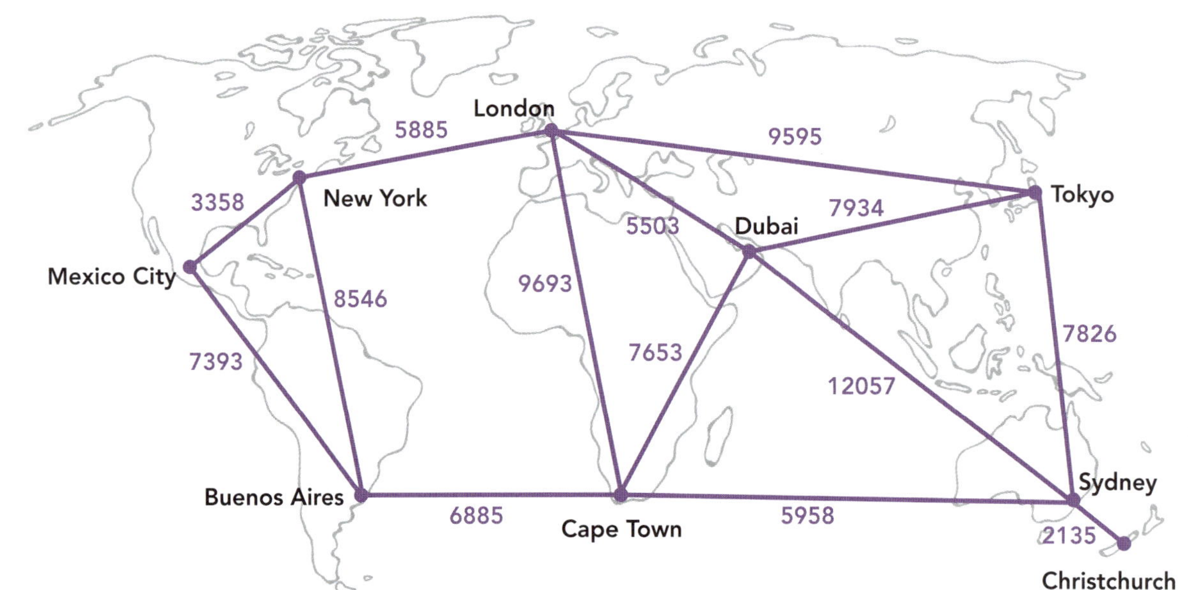

ISBN: 9780170389433

Maximum spanning trees

- A maximum spanning tree connects all the nodes using the **maximum weighting** of edges possible, but using the **minimum number** of edges possible.
- There can be **more than one** maximum spanning tree.
- These are less common than minimum spanning trees.
- Example of use:
  - assessment of routes for maximum scenic values, tourist attractions, etc.

Example:

Step 1: Make an **ordered** list of all the path lengths.

30	23	22	20	19	18	17	16	15

Step 2: Highlight any paths that lead to nodes with the value of 1. You **must** use these.

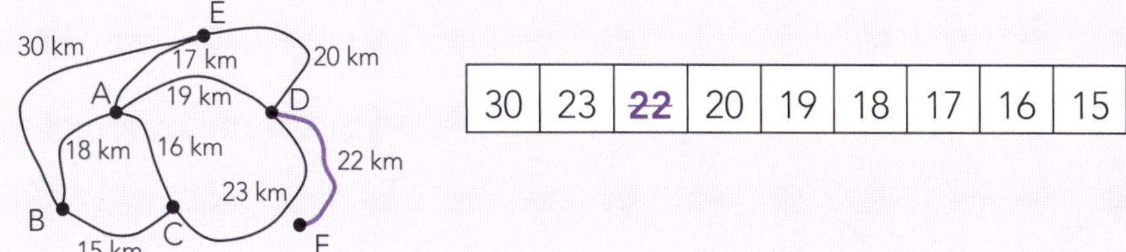

30	23	~~22~~	20	19	18	17	16	15

Step 3: Highlight the longest paths on your network, crossing each one off as you go.

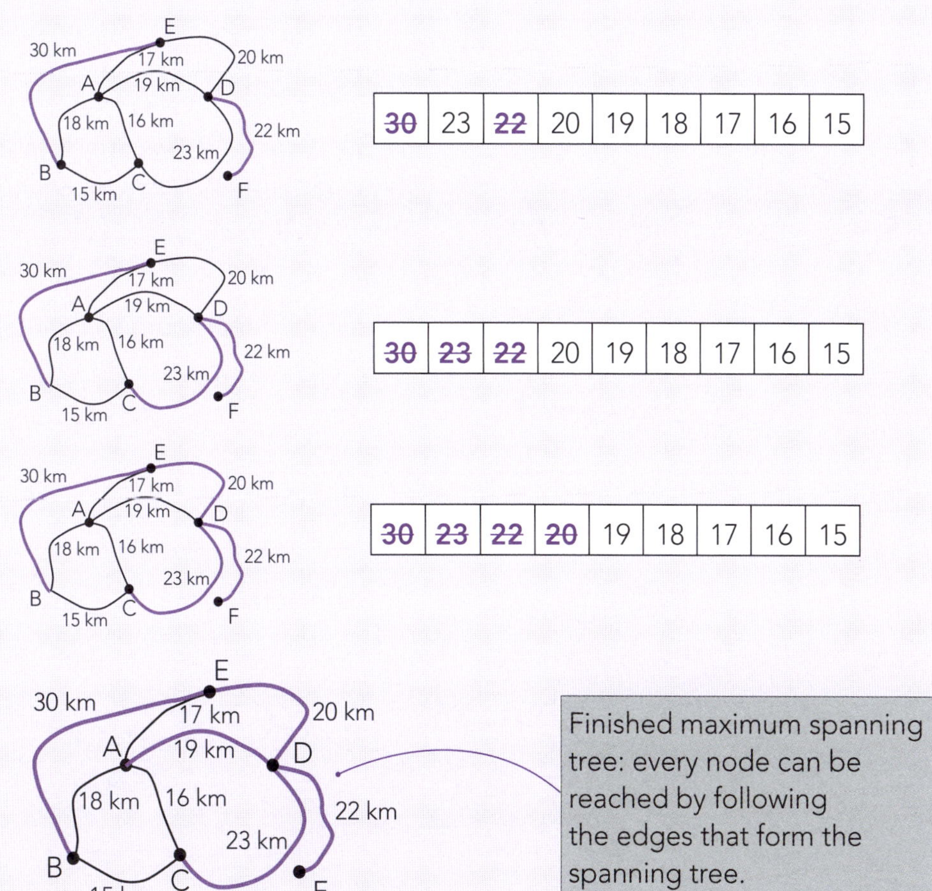

~~30~~	23	~~22~~	20	19	18	17	16	15

~~30~~	~~23~~	~~22~~	20	19	18	17	16	15

~~30~~	~~23~~	~~22~~	~~20~~	19	18	17	16	15

Step 4: Calculate the total length of the roads needed to connect all the towns:
Length = 22 + 30 + 23 + 20 + 19 = **114 km**.

ISBN: 9780170389433

Highlight the maximum spanning tree for each diagram, and give its value, with units.

1 The diagram shows distances between towns.

Length of maximum spanning tree:

____

18	15	13					

2 The diagram shows the times (minutes) taken to walk a series of tracks in a park.

Length of maximum spanning tree:

____

9											

3 The diagram shows the distances (nautical miles) between anchorages in a marine park.

Length of maximum spanning tree:

____

4

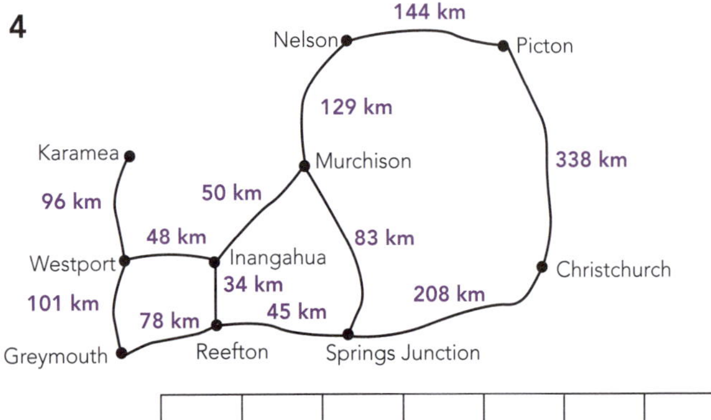

The diagram shows the distances between some South Island towns.

Length of maximum spanning tree:

____

ISBN: 9780170389433

5 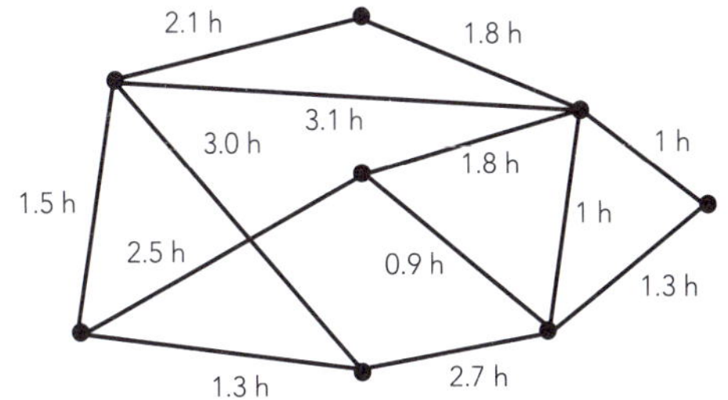

The diagram shows the time needed each week to clear sections of tracks in a national park.

Length of maximum spanning tree:

____

6 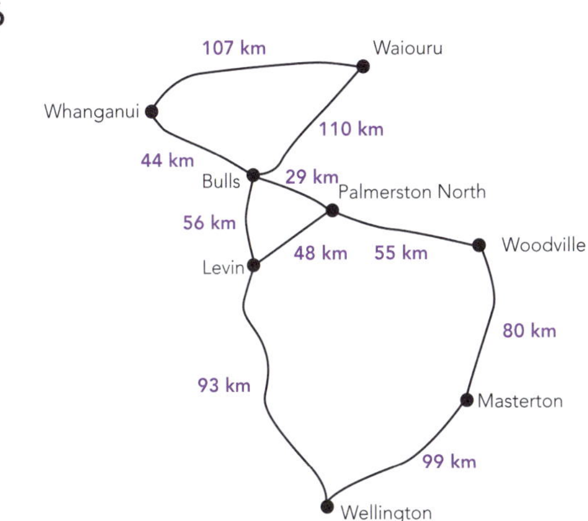

The diagram shows the distances between some North Island towns.

Length of maximum spanning tree:

____

7 The map shows the distances between cities (km). Hugo loves flying (and he also has plenty of money), so he would like to visit every location on the map by travelling the longest distance possible.

a On the map, highlight the longest flights needed to connect every location.

b Calculate its length. ______________________

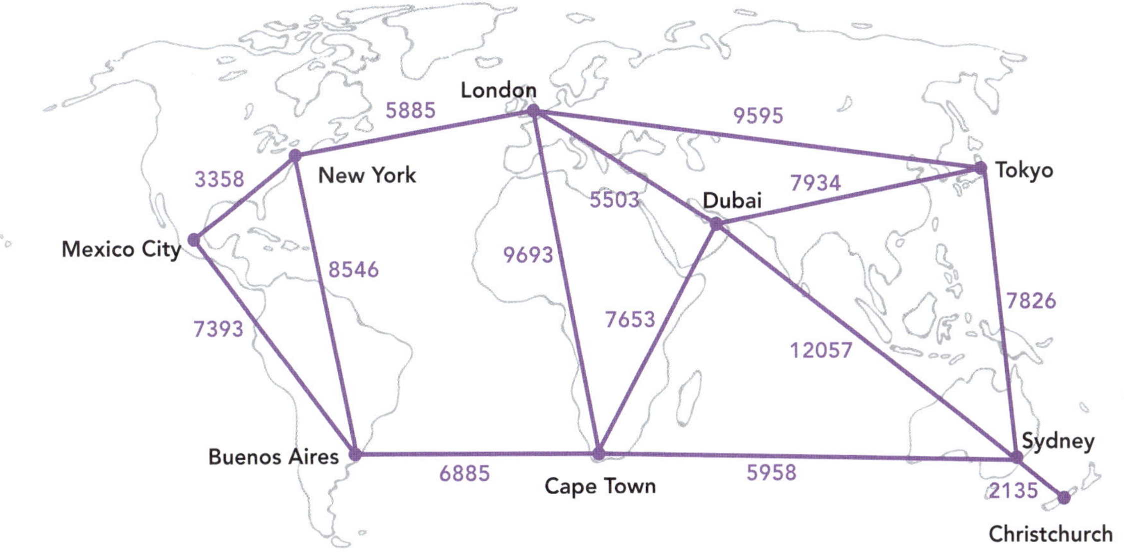

ISBN: 9780170389433

Practice tasks

Practice task one

Xavier, Yelena, Zed and Alfred are going on a road trip.

- Xavier is really eager to get from Fairlie to Invercargill as soon as possible to ensure they are in time for the All Blacks test on Friday night.
- Zed is currently studying networks, and he wants to create a spanning tree which could connect every town exactly once using the roads with as many rest areas as possible.
- Alfred is in the same class, and wants to create a spanning tree which would connect every town exactly once using the shortest road length possible.
- Both Zed and Alfred want to include as many of the roads from their spanning trees as possible into their final route.
- Yelena is into sightseeing, so she wants to visit every town without driving any roads more than once. Is this possible? Explain your answer.

This task requires you to find the optimal path for each passenger, and then design a network that meets as many of their requirements as possible. Present your optimal network, justify your decisions, and make a recommendation for the group.

Resource 1

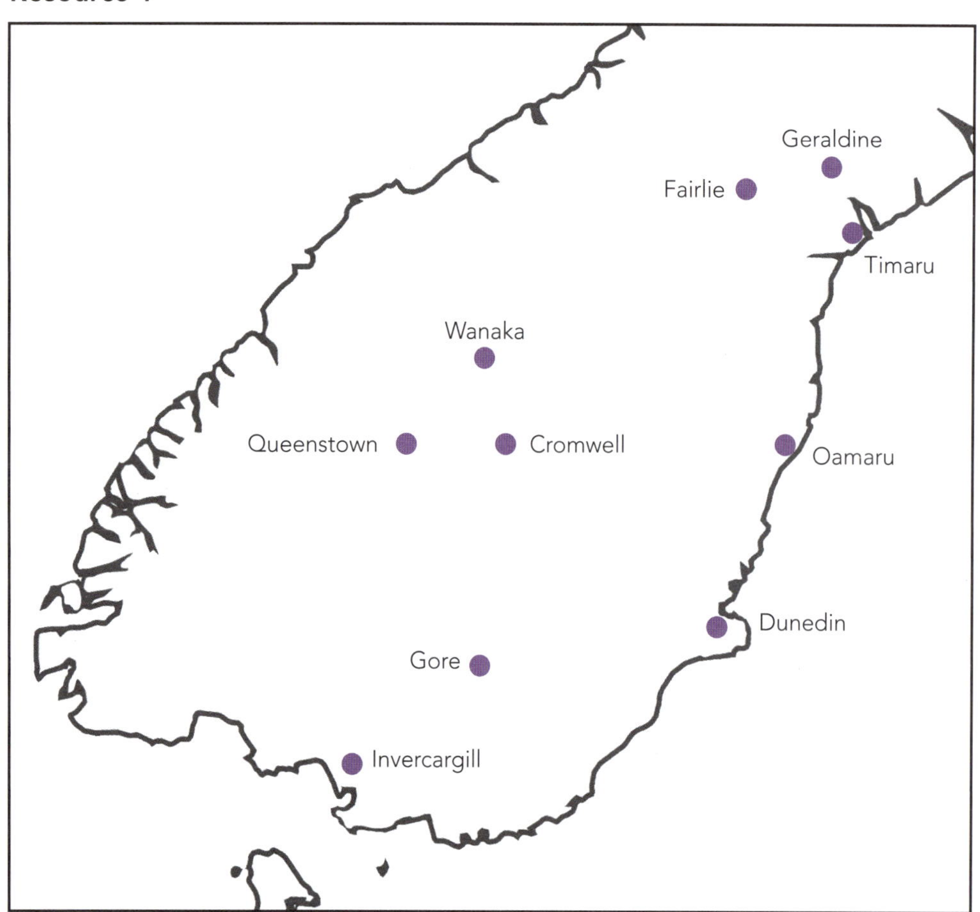

  ISBN: 9780170389433

Resource 2: Distances (km) between towns.

	Geraldine	Fairlie	Timaru	Oamaru	Cromwell	Wanaka	Dunedin	Queenstown	Gore	Invercargill
Geraldine		46	35							
Fairlie	46		63			243				
Timaru	35	63		84						
Oamaru			84			231	115			
Cromwell						55	221	62		
Wanaka		243		231	55					
Dunedin				115	221				151	217
Queenstown					62				169	187
Gore							151	169		66
Invercargill							217	187	66	

Resource 3: The number of rest areas between towns.

	Geraldine	Fairlie	Timaru	Oamaru	Cromwell	Wanaka	Dunedin	Queenstown	Gore	Invercargill
Geraldine		2	1							
Fairlie	2		0			8				
Timaru	1	0		4						
Oamaru			4			9	3			
Cromwell						1	4	2		
Wanaka		8		9	1					
Dunedin				3	4				10	5
Queenstown					2				7	3
Gore							10	7		2
Invercargill							5	3	2	

ISBN: 9780170389433

Xavier

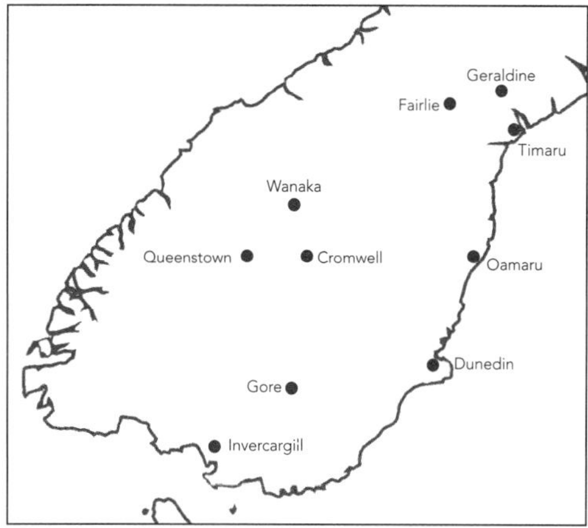

Zed

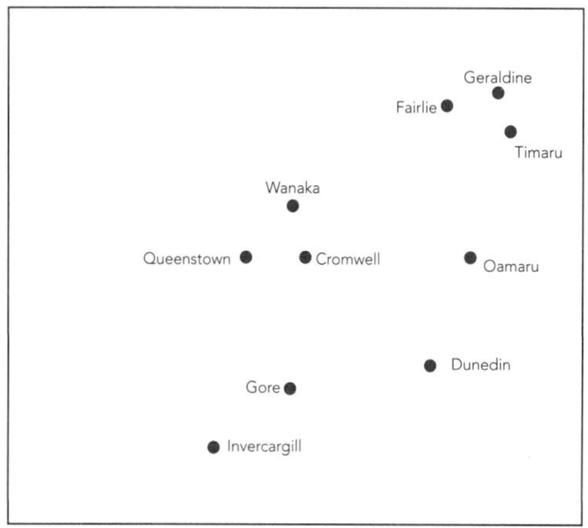

Alfred

ISBN: 9780170389433

Yelena

Optimal network

Practice task two

Bella, Aroha, Eli and Jack are staying in the camping ground near their favourite beach.

- Bella and Aroha have walked to the playground, and they arrange to meet at the jumping bridge later in the afternoon.
- Bella is feeling tired after the day's travel, so she wants to walk from the playground to the jumping bridge via the shortest possible route. Find a route for her, and calculate its length.
- Aroha is training for rugby, so she wants to run from the playground to the jumping bridge via the longest route possible without back-tracking. She would also like to visit every location (including the motels) exactly once on the way. She is happy to run on both the roads and the tracks. Find a route for her, and calculate its length.
- Jack wants to leave from the camping ground and visit each location on the **tracks,** except the mud slide and the motels. However, he is feeling lazy so, although he is happy to revisit a location, he wants to walk every section of the track exactly once. He does not want to walk on the road. Can he do it? Explain your answer.
- Eli works for the local council. They would like to install lighting along the tracks (not the road). They cannot afford to light all the tracks, but they want to light the minimum required so that a person can walk to every location, except the jumping bridge, using a lighted track. Installing the lighting using a machine to bury the wiring in a trench will cost $65 per metre. Eli has the job of finding a way to install the lighting as cheaply as possible. Find the best route for the lighting, and the cost of installation.

Unfortunately for the council, they discover that they are not allowed to use a machine to bury the wiring **within 10 m** of the giant totara. There are two options available to them:

A: Use Eli's route but stop using the digging machine 10 m away from the giant totara, and hand-dig trenches around the tree to connect up the lighting. This will involve 38 m of hand-digging at a cost of $240 per m.

B: Find a different route from Eli's one. This route would have just one lit path leading to the giant totara, but the lighting would stop 10 m away from the tree.

Investigate options A and B and make a recommendation to the council.

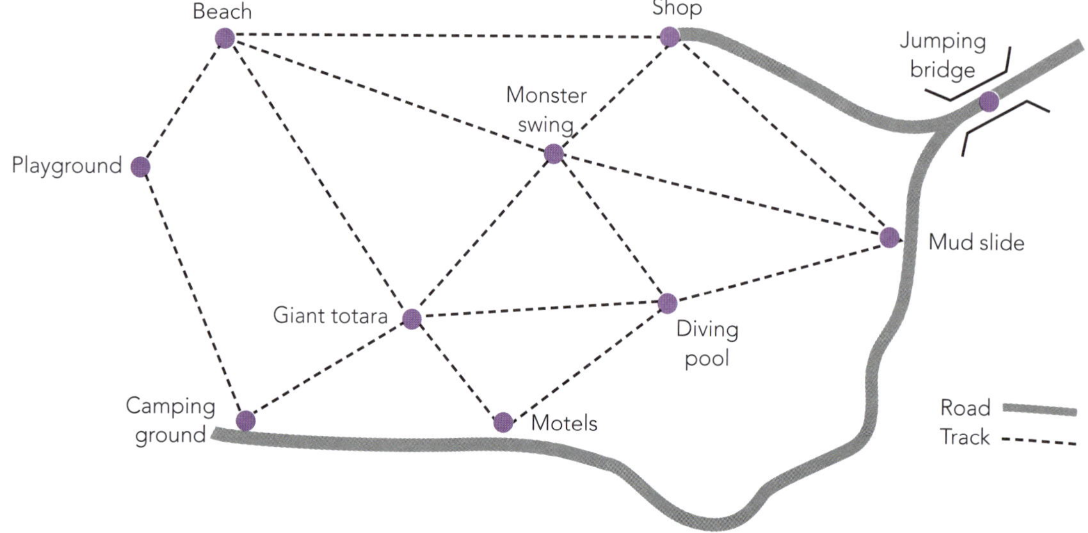

ISBN: 9780170389433

Distances between locations (m):

	Beach	Playground	Camping ground	Giant totara	Motels	Monster swing	Shop	Mud slide	Diving pool	Jumping bridge
Beach		18		123		211	280			
Playground	18		96							
Camping ground		96		83	125 (on road)					
Giant totara	123		83		52	109			132	
Motels			125 (on road)	52				350 (on road)	92	
Monster swing	211			109			85	95	90	
Shop	280					85		98		130 (on road)
Mud slide					350 (on road)	95	98		110	103 (on road)
Diving pool				132	92	90		110		
Jumping bridge							130 (on road)	103 (on road)		

Bella

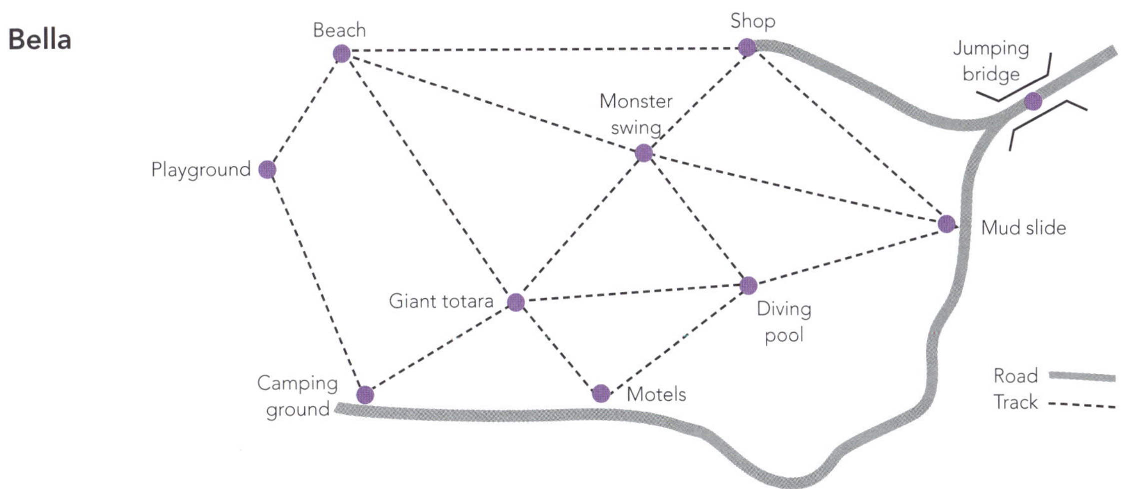

Aroha

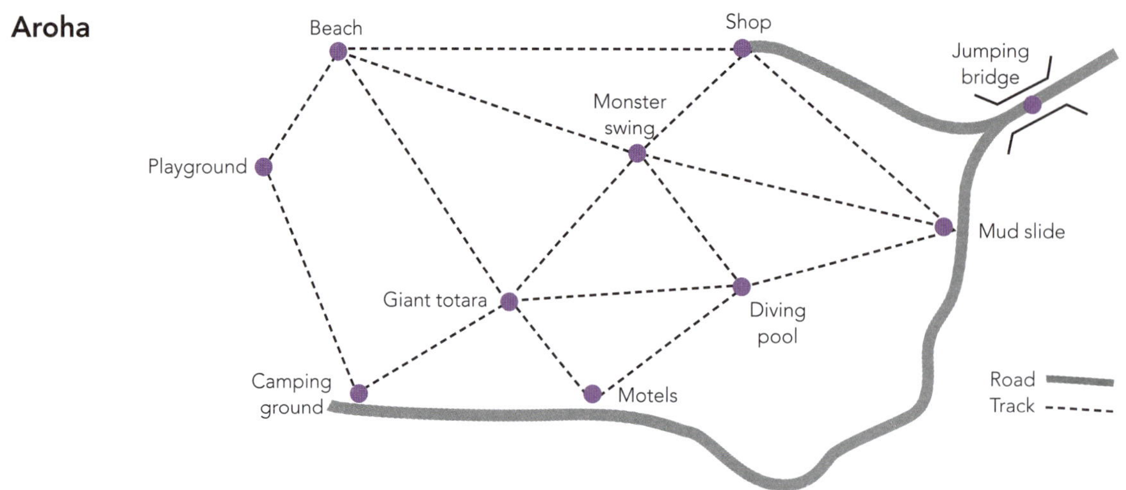

Jack

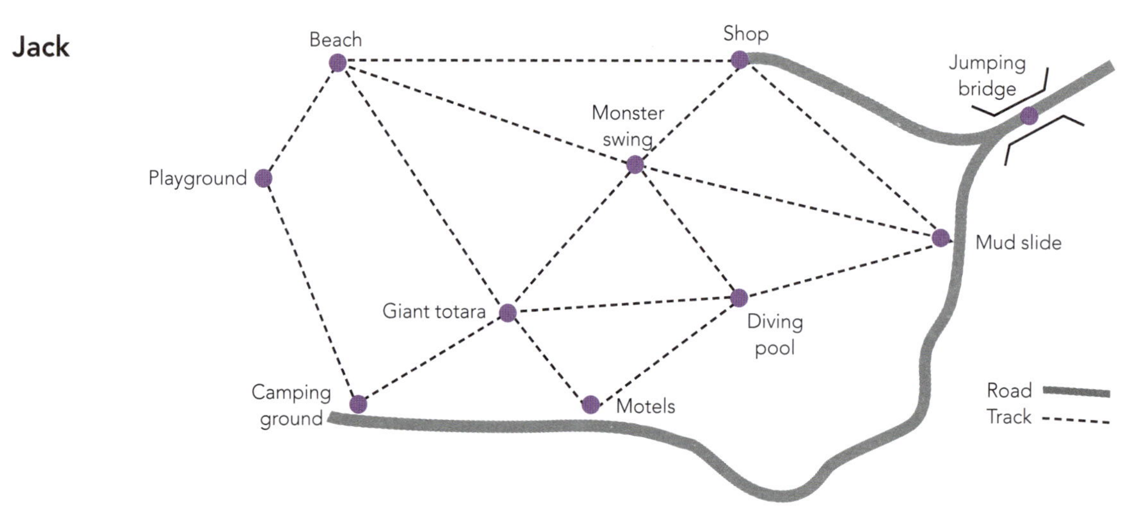

ISBN: 9780170389433

Eli

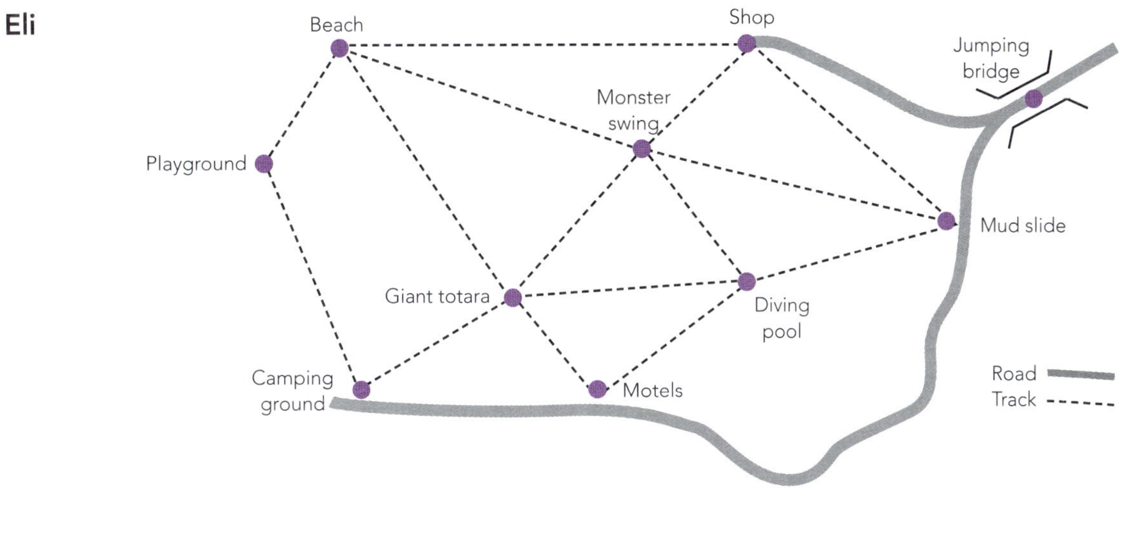

Options A and B and recommendation to the council:

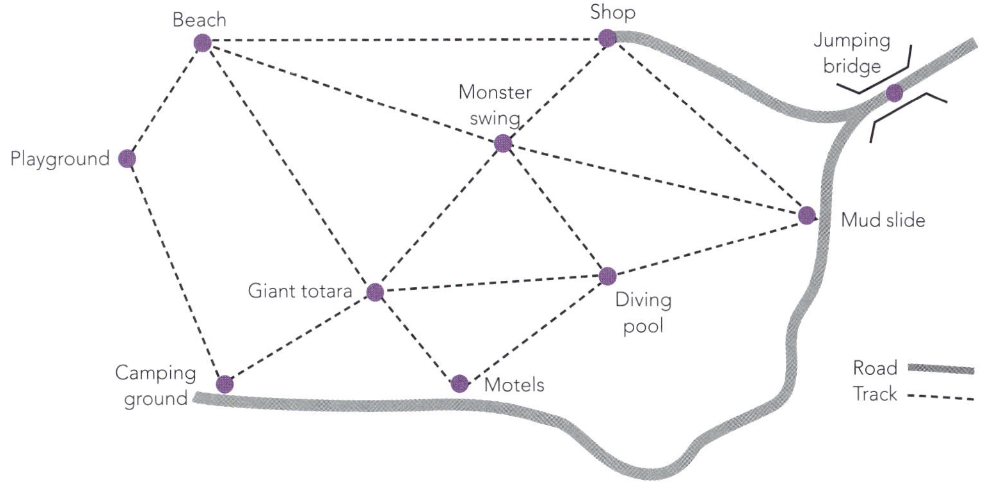

Practice task three

Santa Town has expanded over recent years, but not in a very logical way, so Santa has some major logistical problems.

- The elves have to transport the toys from the factory to the wrapping room. Find the shortest route for them and find its length.
- Santa is getting a little pudgy. In order to get some exercise each day, he would like to ski every track in the town exactly once. Is this possible? If so, explain your answer, and describe possible routes for Santa and the reasons for them.
- Christmas is approaching, and greater efficiency is needed. Santa decides that a monorail connecting all the locations would help. However, at $450 per metre, monorails are very expensive, and he would like to connect every location using the shortest route possible. Find the shortest route for the monorail, its total length and its cost. Call this route A.
- There is only a 39 m gap between Mrs Claus's candy kitchen and the sleigh servicing shop. There is no track at all over the 39 m, and the terrain would be particularly difficult for construction of a monorail connection. Investigate the route and the minimum length of monorail that would be required if this connection could be made. Call this route B.
- Assume that all monorail sections cost $450 per metre except for this 39 m length. Calculate the maximum cost per metre of monorail needed to connect across the 39 m gap that would make Route B a cheaper option.

Resource 1: The map shows the not-very-sensible layout of Santa Town at the North Pole, with the routes between each location.

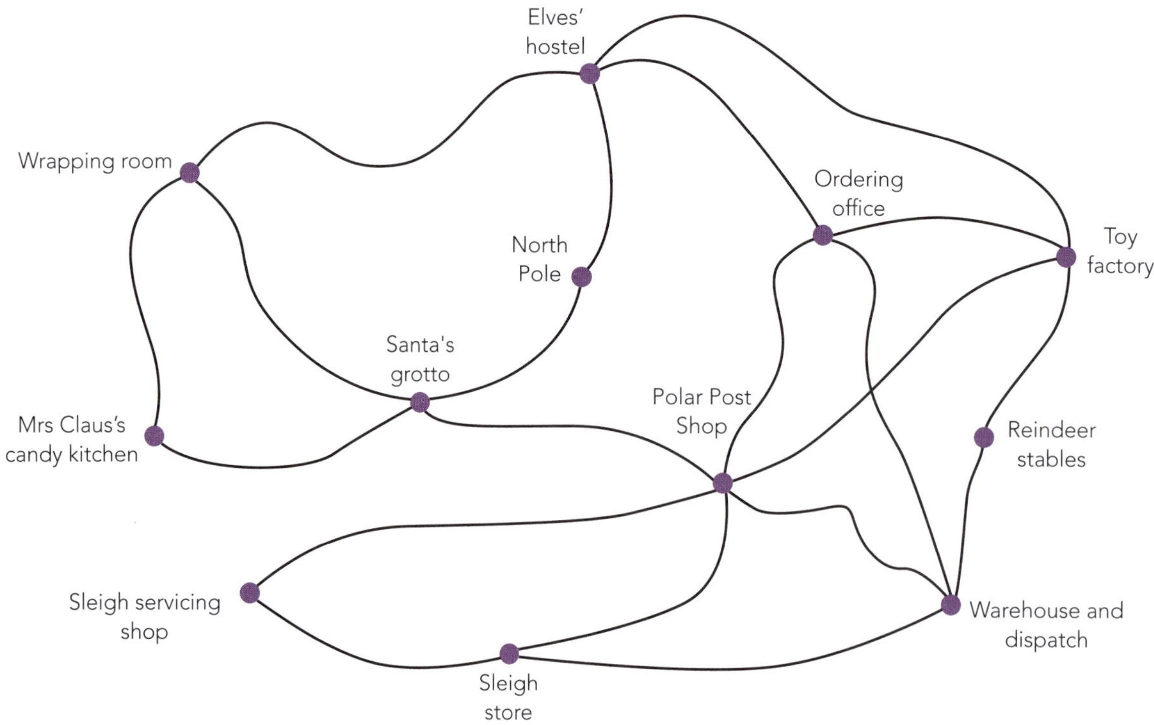

ISBN: 9780170389433

Resource 2: The table shows the distances (m) between each location.

	North Pole	Elves' hostel	Ordering office	Toy factory	Warehouse and dispatch	Polar Post Shop	Sleigh store	Santa's grotto	Wrapping room	Mrs Claus's candy kitchen	Reindeer stables	Sleigh servicing shop
North Pole												
Elves' hostel	69											
Ordering office		142										
Toy factory		176	81									
Warehouse and dispatch			228									
Polar Post Shop			97	183	57							
Sleigh store					271	171						
Santa's grotto	77					124						
Wrapping room		195						116				
Mrs Claus's candy kitchen								84	83			
Reindeer stables				64	56							
Sleigh servicing shop						305	128					

Elves' transport of toys from the factory to the wrapping room

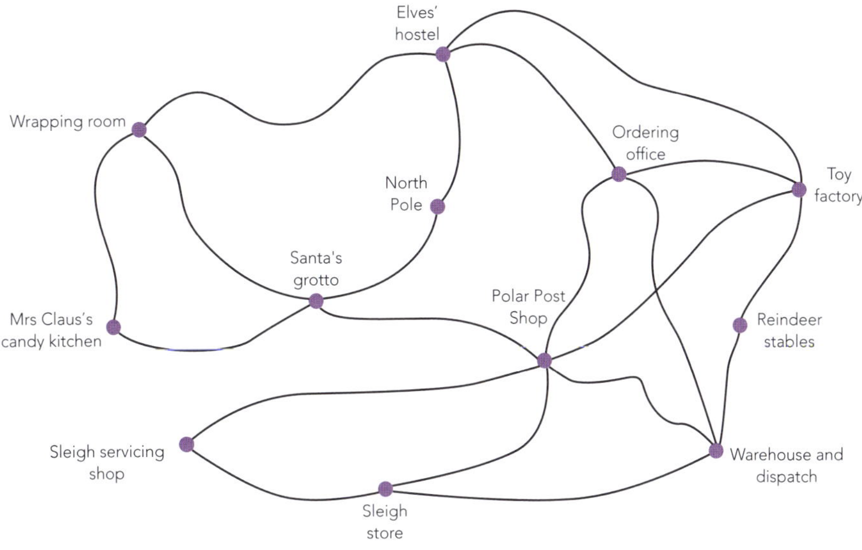

ISBN: 9780170389433

Santa's skiing route

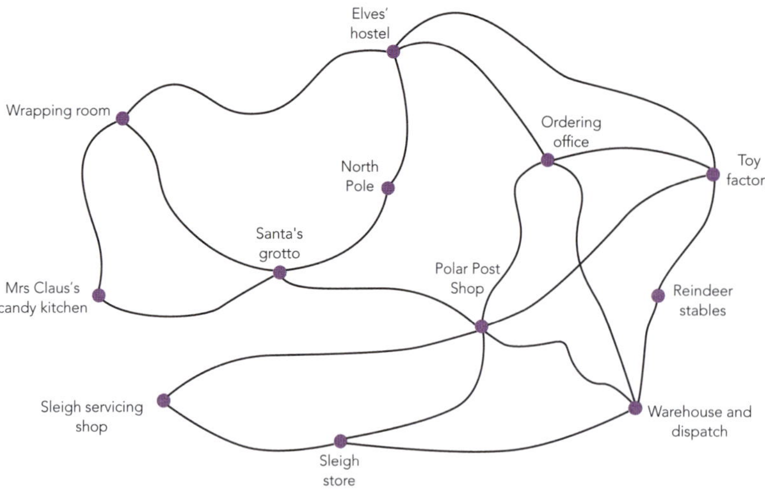

Route A for the monorail

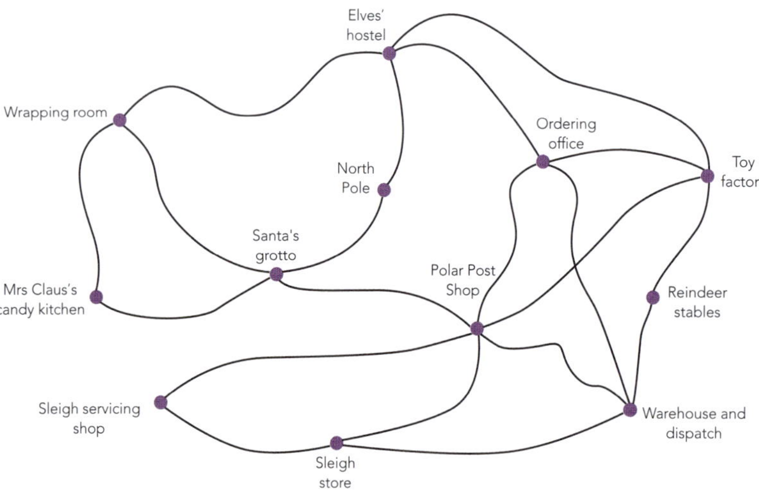

ISBN: 9780170389433

Route B for the monorail

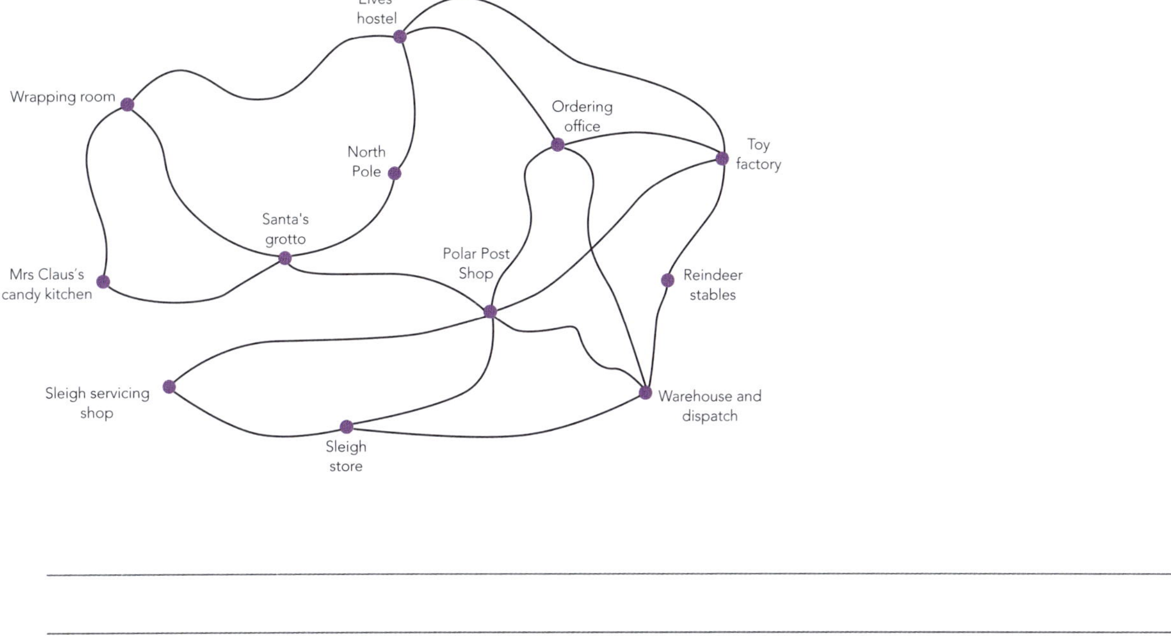

Answers

Nodes and edges (pp. 6–7)

1 Nodes: 4
Edges: 7
2 Nodes: 5
Edges: 7
3 Nodes: 4
Edges: 6
4 Nodes: 6
Edges: 12
5 Nodes: 13
Edges: 18
6 Nodes: 14
Edges: 19

The degree of a node (p. 8)

1

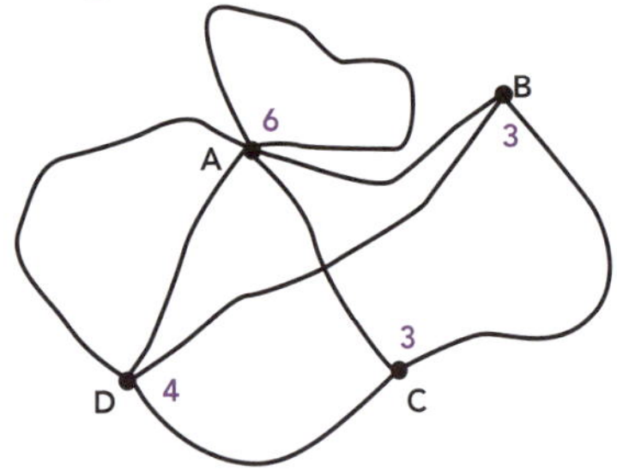

Node	A	B	C	D
Degree	6	3	3	4

2

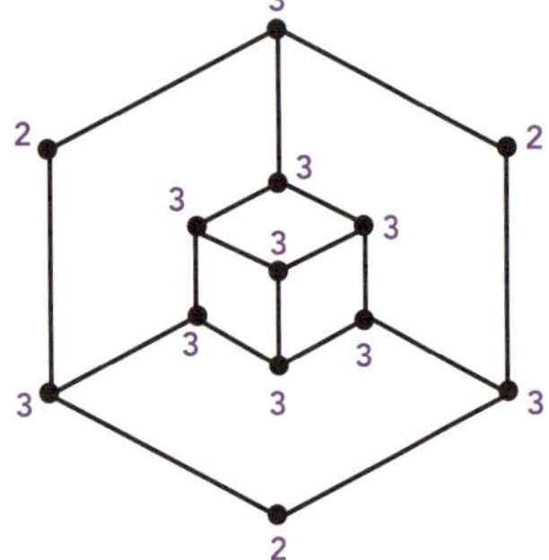

3

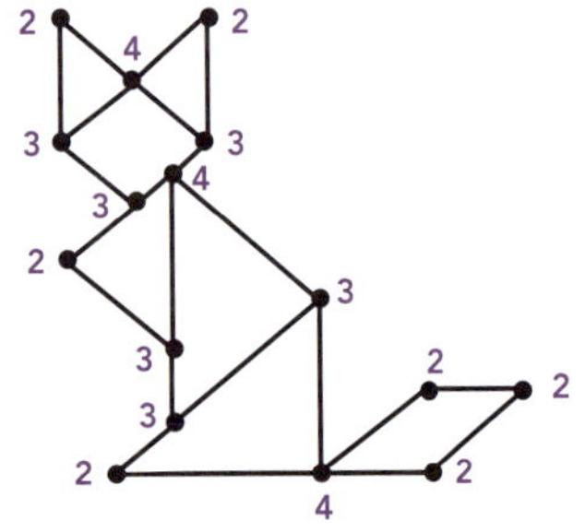

4

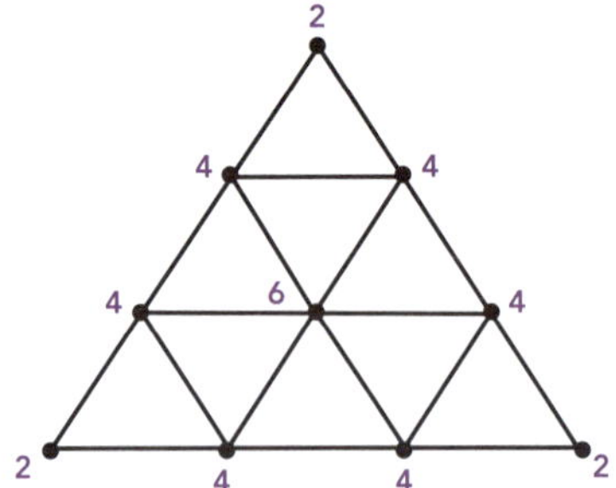

Odd and even nodes (pp. 9–10)

1

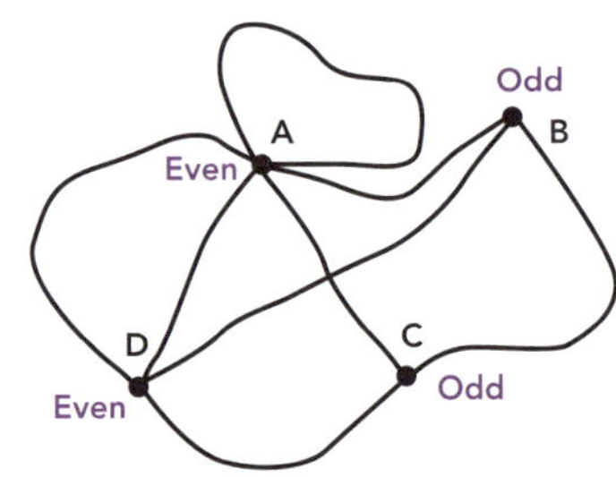

Node	A	B	C	D
Degree	6	3	3	4
Odd/ Even	Even	Odd	Odd	Even

2

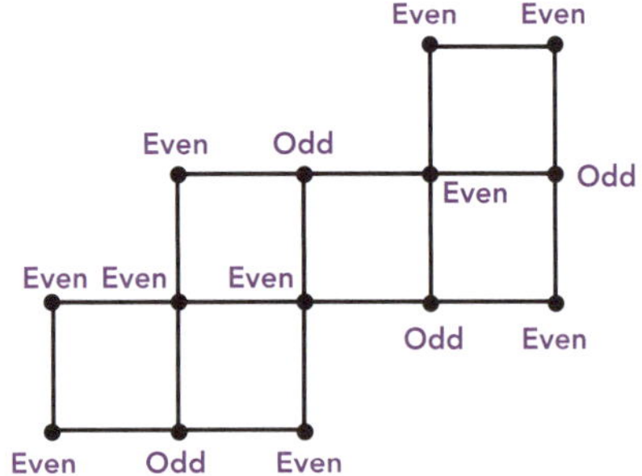

3

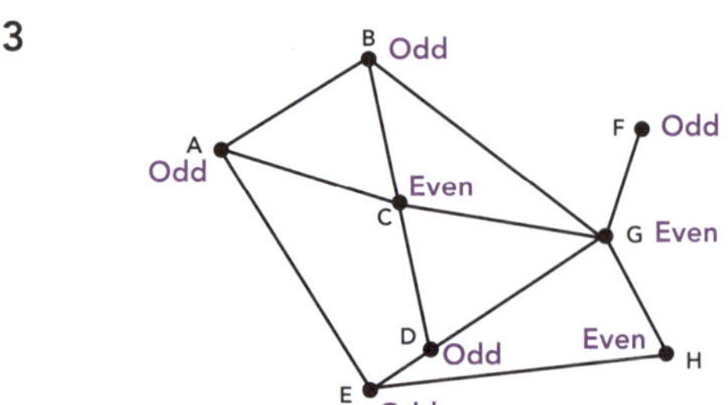

Node	A	B	C	D	E	F	G	H
Degree	3	3	4	3	3	1	5	2
Odd/ Even	Odd	Odd	Even	Odd	Odd	Odd	Odd	Even

ISBN: 9780170389433

4

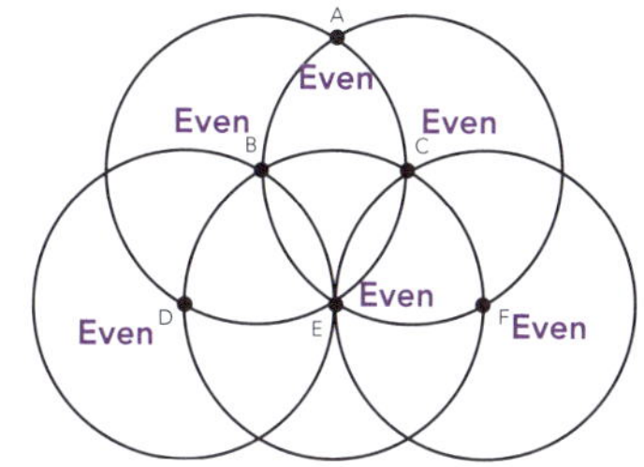

Node	A	B	C	D	E	F
Degree	4	6	6	4	8	4
Odd/ Even	Even	Even	Even	Even	Even	Even

5

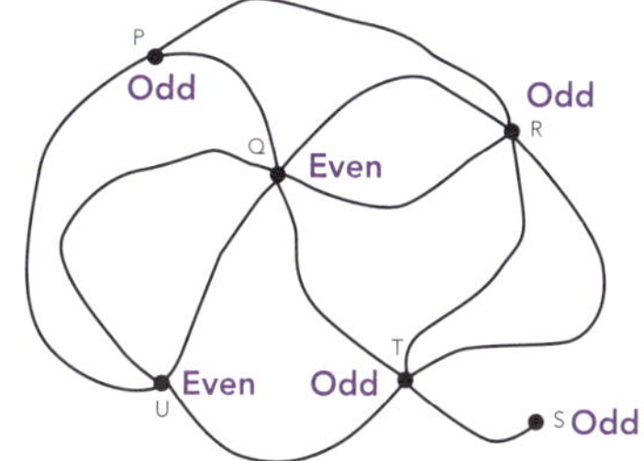

Node	P	Q	R	S	T	U
Degree	3	6	5	1	5	4
Odd/ Even	Odd	Even	Odd	Odd	Odd	Even

6

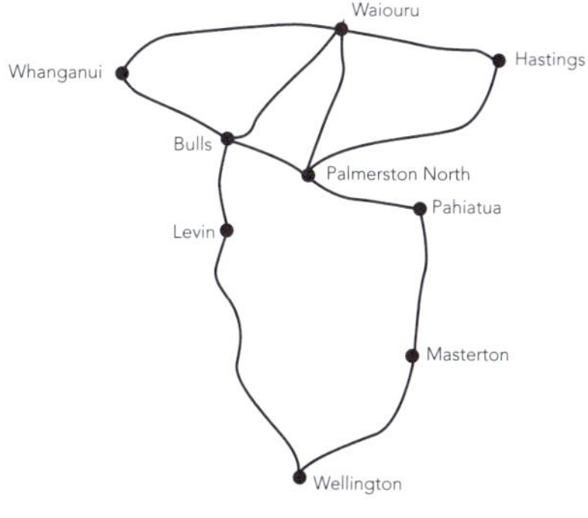

Node	Degree	Odd/Even
Whanganui	2	Even
Bulls	4	Even
Waiouru	4	Even
Hastings	2	Even
Palmerston North	4	Even
Pahiatua	2	Even
Levin	2	Even
Wellington	2	Even
Masterton	2	Even

Paths and circuits (pp. 11–12)

1 Path: 75
2 Circuit: 12
3 Path: 66
4 Circuit: 9
5 Path: 28
6 Path: 130

Constructing networks (pp. 13–19)

(Check with your teacher if you have drawn a different network.)

1 a

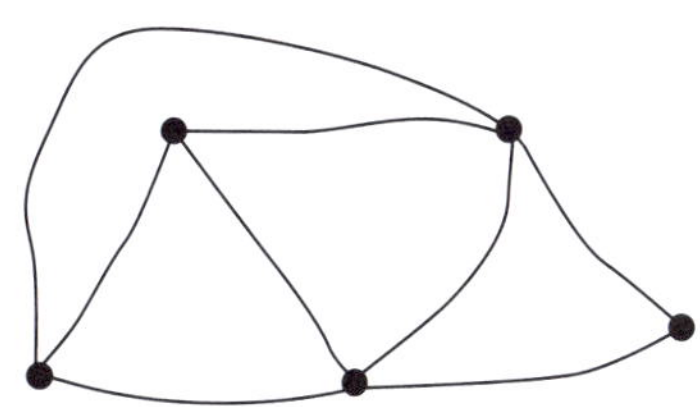

b

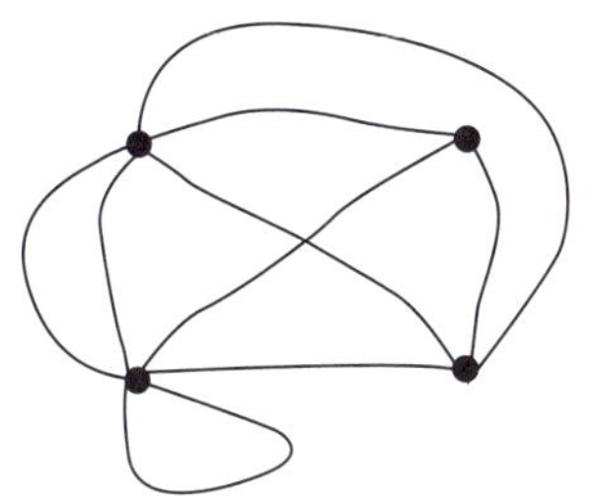

c

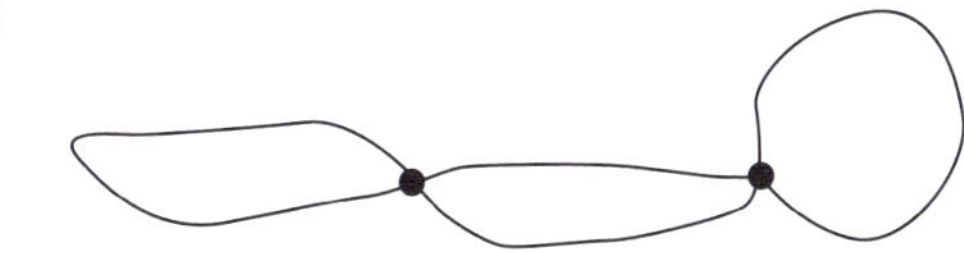

d

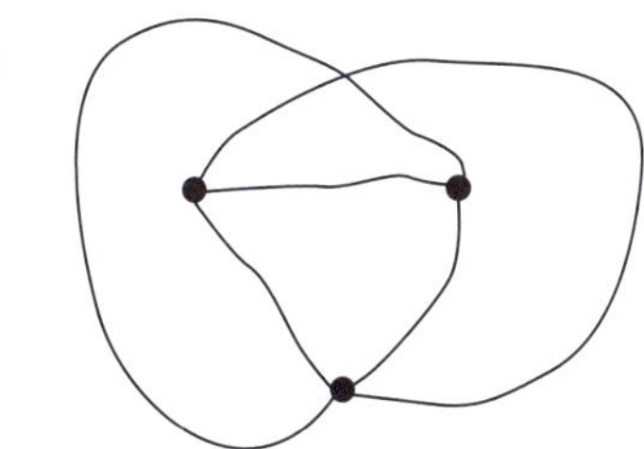

e

f Not possible

g

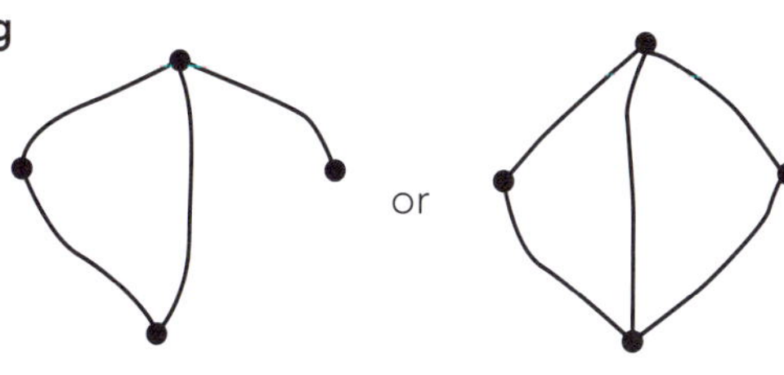

h Not possible

i Not possible

j

or

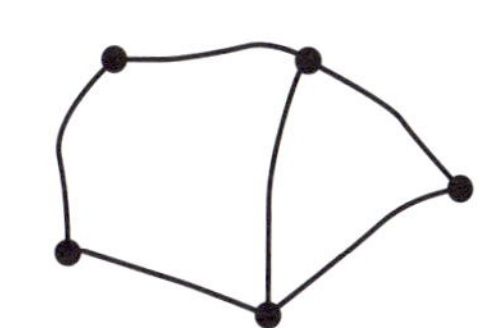

ISBN: 9780170389433

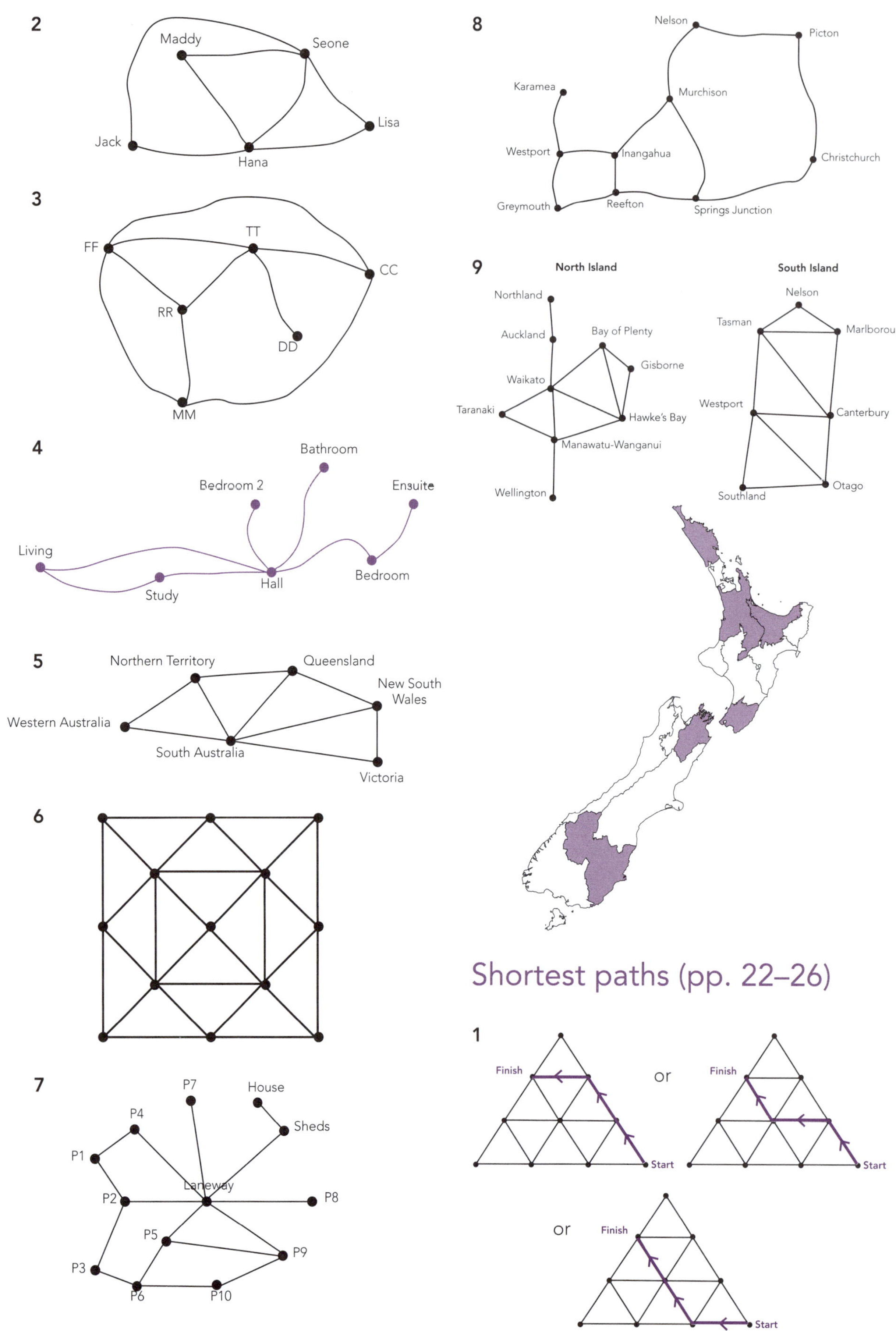

Shortest paths (pp. 22–26)

1

 ISBN: 9780170389433

2 Shortest path length: 11 km

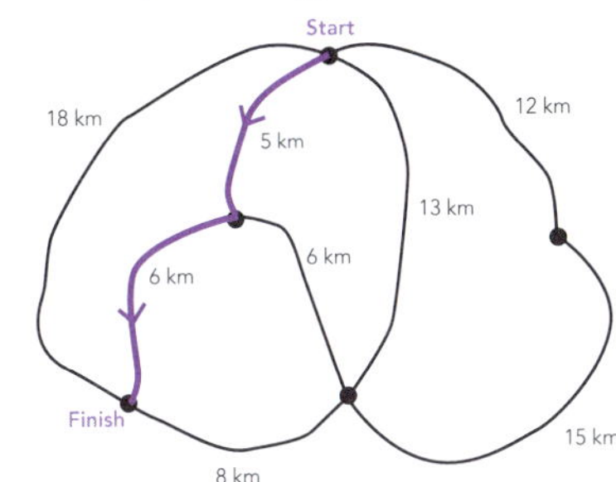

3 Shortest path length: 5.3 hours (two correct routes)

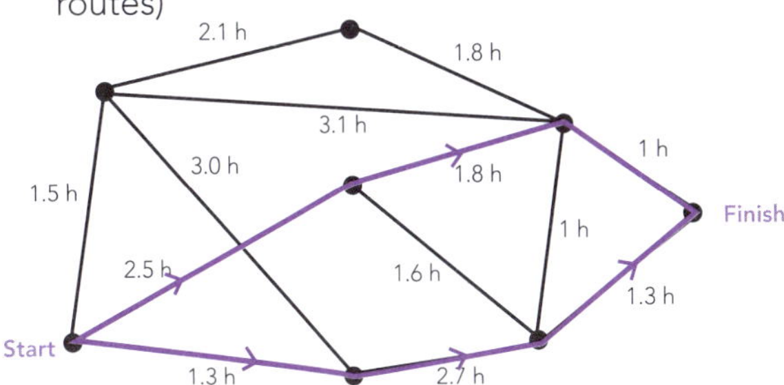

4 Shortest path length: 10 minutes

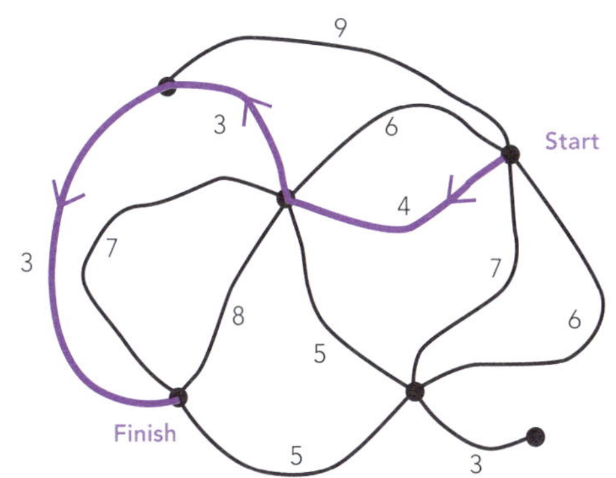

5 **a** Shortest path from the start (V) to finish (Z): V → T → S → R → Z
Length of shortest path from the start (V) to finish (Z): 104 m

b Shortest path from the start (V) to finish (M): V → T → P → Q → M
Length of shortest path from the start (V) to finish (M): 105 m

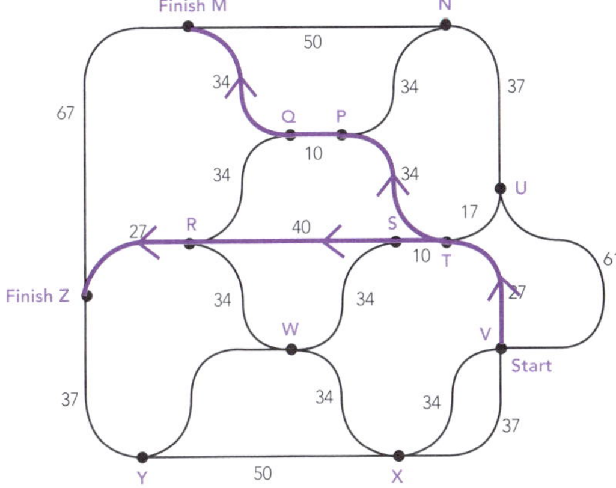

6 **a** Greymouth to Murchison description: Greymouth → Reefton → Inangahua → Murchison
Shortest path from Greymouth to Murchison: 162 km

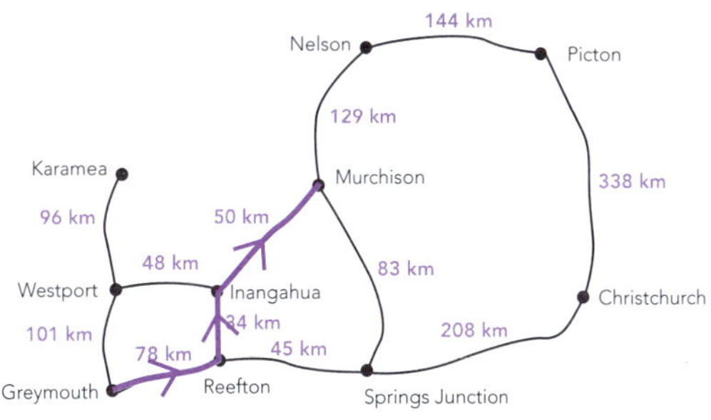

b Greymouth to Picton description: Greymouth → Reefton → Inangahua → Murchison → Nelson → Picton
Shortest path from Greymouth to Picton: 435 km

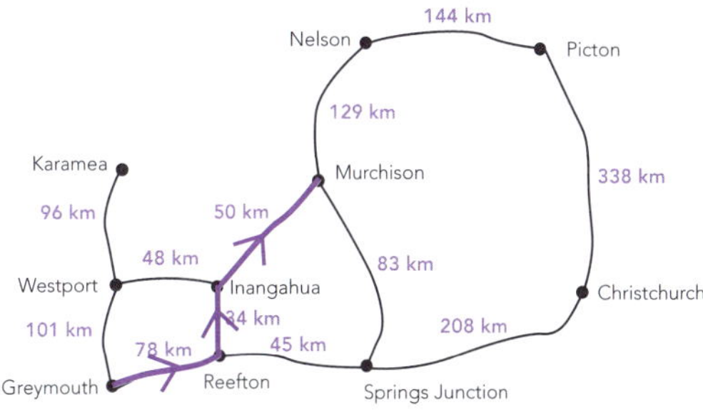

7 **a** Shortest path from London to Sydney: London → Cape Town → Sydney
Length: 15 651 km

b Shortest path from Cape Town to New York: Cape Town → Buenos Aires → New York
Length: 15 431 km

c Shortest path from Cape Town to Tokyo: Cape Town → Sydney → Tokyo
Length: 13 784 km

d Shortest path from Buenos Aires to Tokyo: Buenos Aires → Cape Town → Sydney → Tokyo
Length: 20 669 km

e Shortest path from Mexico City to Sydney: Mexico City → Buenos Aires → Cape Town → Sydney
Length: 20 236 km

ISBN: 9780170389433

Traversability (pp. 27–35)

1

Node	A	B	C	D	E
Degree	3	3	2	4	2
Odd/ Even	Odd	Odd	Even	Even	Even

Traversable? Yes, Euler path.
Reason: Exactly two odd nodes (A and B).

2

Node	A	B	C	D
Degree	4	3	3	4
Odd/ Even	Even	Odd	Odd	Even

Traversable? Yes, Euler path.
Reason: Exactly two odd nodes (B and C).

3

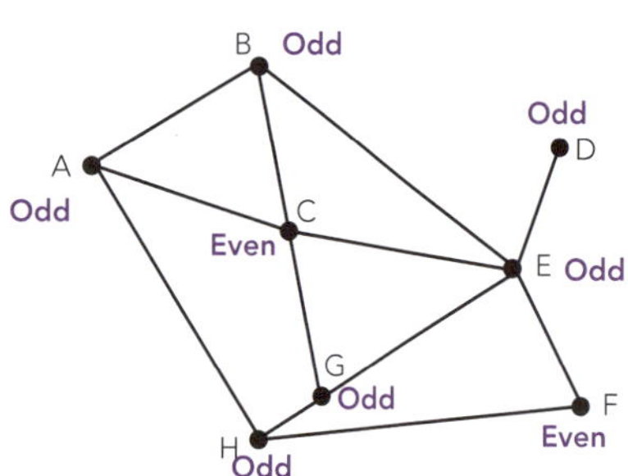

Node	A	B	C	D	E	F	G	H
Degree	3	3	4	1	5	2	3	3
Odd/ Even	Odd	Odd	Even	Odd	Odd	Even	Odd	Odd

Traversable? No.
Reason: More than two odd nodes (A, B, D, E, G and H).

4

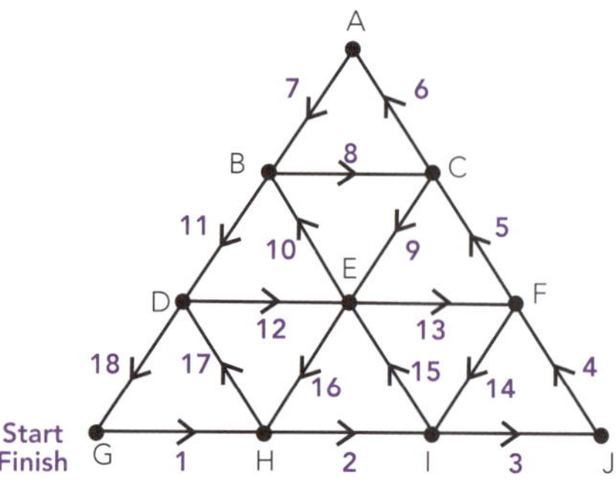

Traversable? Yes, Euler circuit.
Reason: No odd nodes. All nodes are degree two, four or six.

5

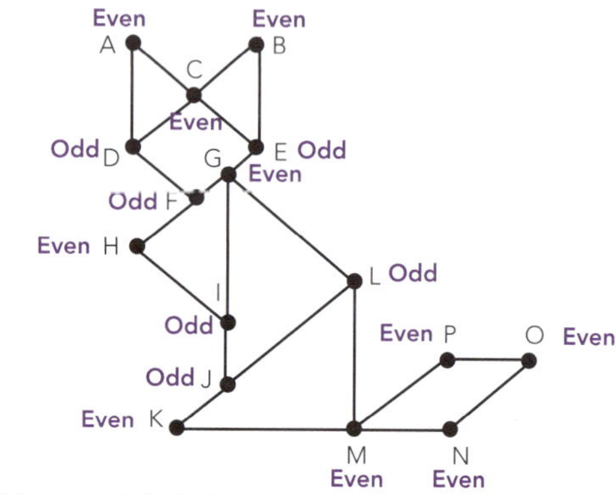

Traversable? No.
Reason: More than two odd nodes (D, E, F, L, I and J) are odd.

6

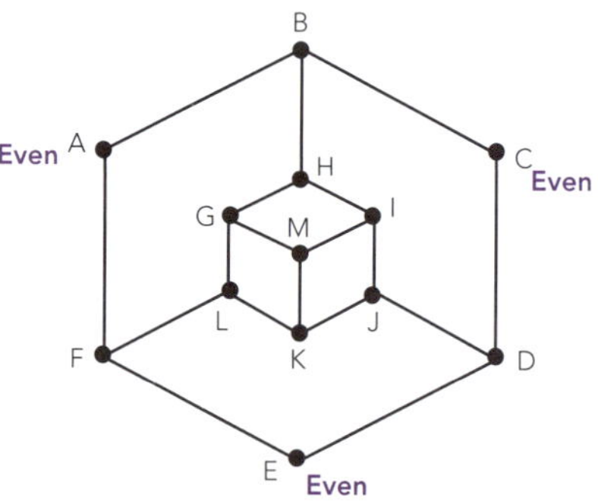

Traversable? No.
Reason:More than two odd nodes. All the nodes except for A, C and E are odd.

7

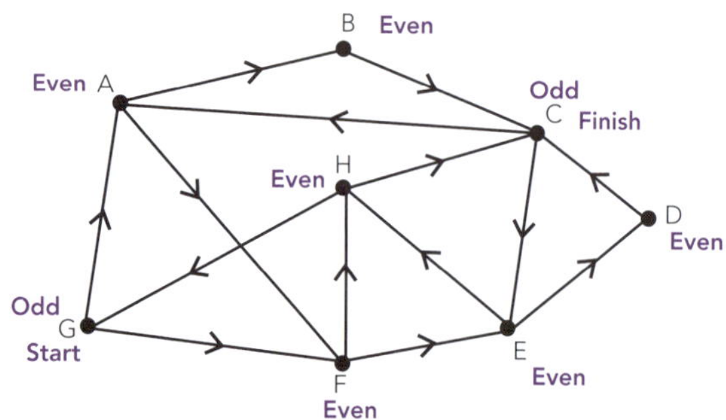

Traversable? Yes, Euler path.
Reason: Exactly two odd nodes (G and C). The rest are even.

ISBN: 9780170389433

8 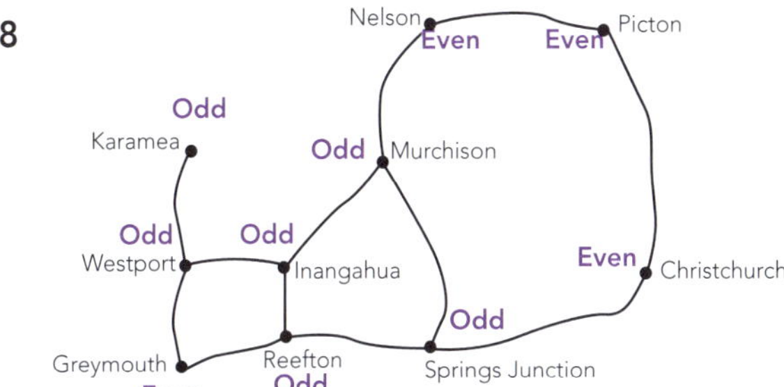

Traversable? No.
Reason: Six odd nodes (Karamea, Westport, Inangahua, Reefton, Springs Junction and Murchison).

9 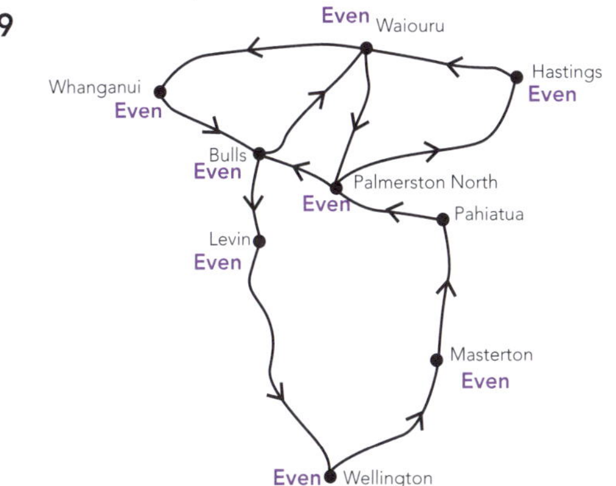

Traversable? Yes, Euler circuit.
Reason: No odd nodes.

10 a Traversable? No. All the nodes are odd.
b Traversable? No. All the nodes are odd.
c Traversable? Yes. All the nodes are even.
d Traversable? No. Five nodes are odd.
e Traversable? No. All the nodes are odd.

Putting it all together (pp. 32–35)

1 a Route: New Caledonia → Fiji → Tonga → Cook Islands
Total flying time: 6h 10m

b

Node	Degree	Odd/Even
Vanuatu	2	Even
Auckland	6	Even
New Caledonia	2	Even
Fiji	6	Even
Tonga	4	Even
Tokelau	2	Even
Cook Islands	4	Even
Samoa	4	Even

All the nodes are even which means the network is traversable, and there will be an Euler circuit which uses every edge exactly once. Therefore the chief executive could fly every route exactly once and start and finish his trip in Fiji.
Example of a route: Fiji → Vanuatu → Auckland → New Caledonia → Fiji → Tokelau → Cook Islands → Samoa → Tonga → Cook Islands → Auckland → Tonga → Fiji → Auckland → Samoa → Fiji.

2 a Either: Gymnasium → Science → Administration Block → Fine Arts → Performing Arts
Or: Gymnasium → Science → Administration Block → Performing Arts
Time required for both routes: 2.7 minutes

b No she cannot make it in time.
Shortest route: Mathematics → Administration Block → Fine Arts → Marae → Library
Time required = 6.5 minutes, so she will be 1.5 minutes late.

c It is not possible to start and finish his walkabout at the Administration Block, because that would be a traversable Euler circuit which requires no odd nodes. The network is not traversable because it has four odd nodes (Science, Languages Technology and the Library).

d There are several options for this answer – if yours is different, check it with your teacher.

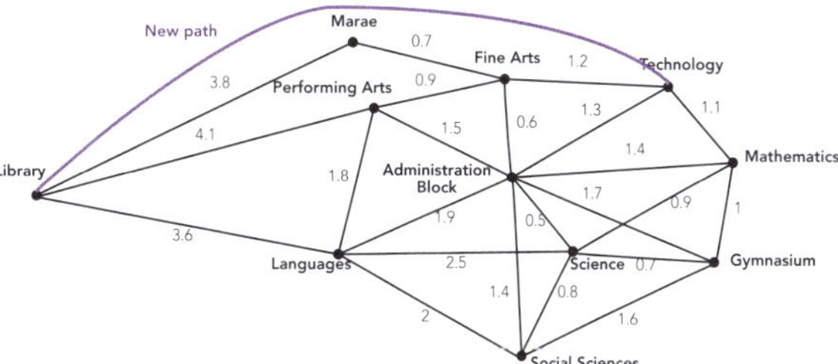

If he built this path, the network would become traversable because there would be exactly two odd nodes, at Science and Languages. However, it would be only traversable via an Euler path, which means he would have to either start at Science and finish at Languages, or the other way around. So, from his office he would need to repeat the paths from the Administration Block to Science and from the Administration Block to Languages.

Networks from tables (pp. 36–42)

1

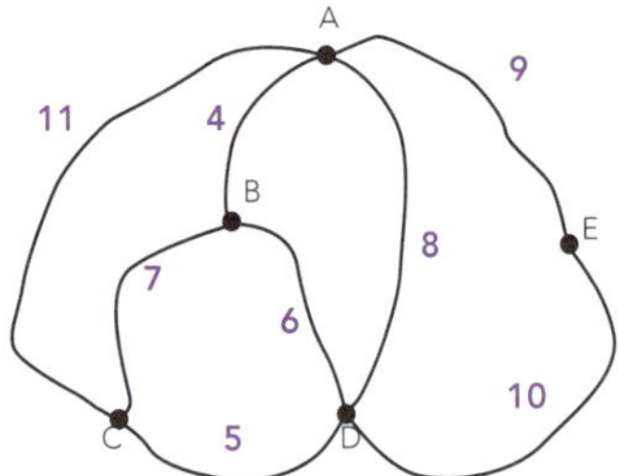

2

A			
21	B		
29	33	C	
41	49	27	D

3

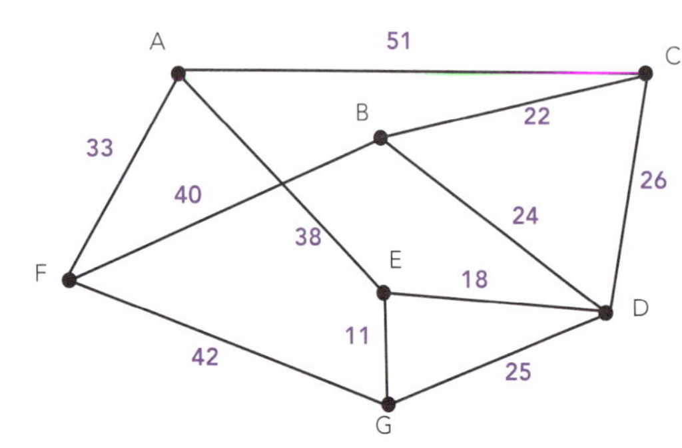

4

	A	B	C	D	E	F	G	H
A		8			16		7	
B	8		15				10	
C		15		13				3
D			13		6			
E	16			6		5	12	11
F					5			
G	7	10			12			9
H			3		11		9	

5

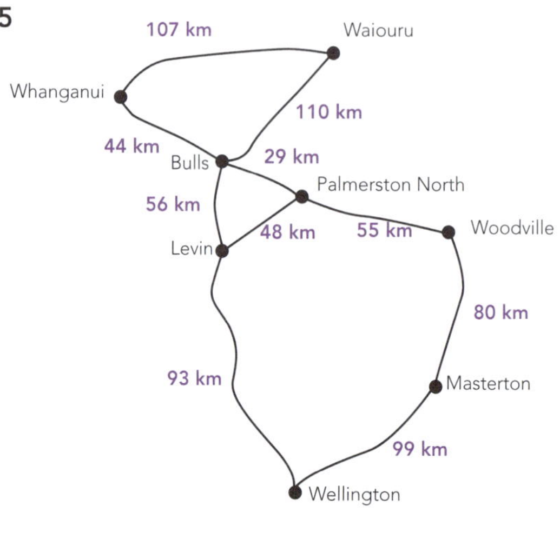

6 a

	Mexico City	Buenos Aires	New York	London	Cape Town	Dubai	Tokyo	Sydney	Christchurch
Mexico City		7393	3358						
Buenos Aires	7393		8546		6885				
New York	3358	8546		5885					
London			5585		9693	5503	9595		
Cape Town		6885		9693		7653		5958	
Dubai				5503	7653		7934	12057	
Tokyo				9595		7934		7826	
Sydney					5958	12057	7826		2135
Christchurch								2135	

b

ISBN: 9780170389433

Spanning trees (pp. 43–49)

Minimum spanning trees (pp. 43–46)

1 Length of minimum spanning tree: 29 km

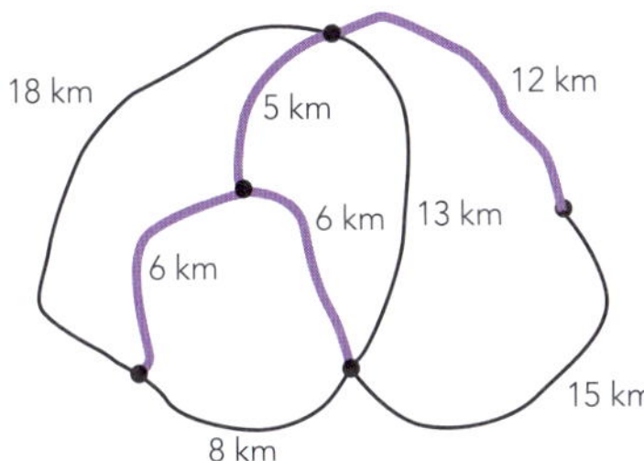

2 Length of minimum spanning tree: 18 minutes

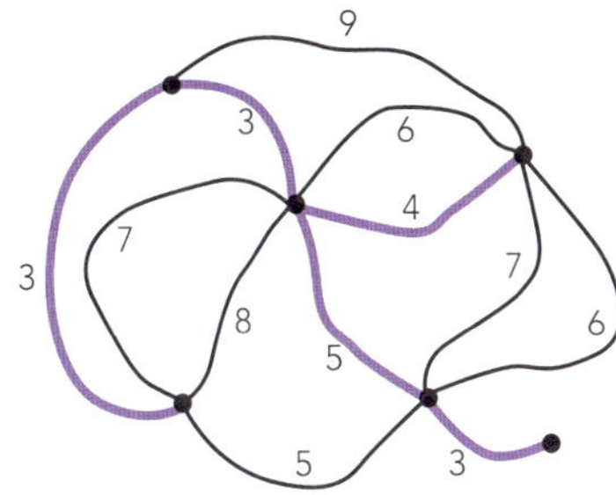

3 Length of minimum spanning tree: 59 nautical miles

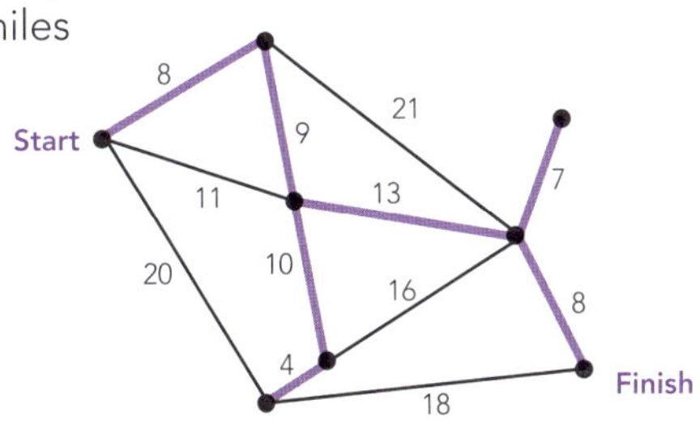

4 Length of minimum spanning tree: 832 km

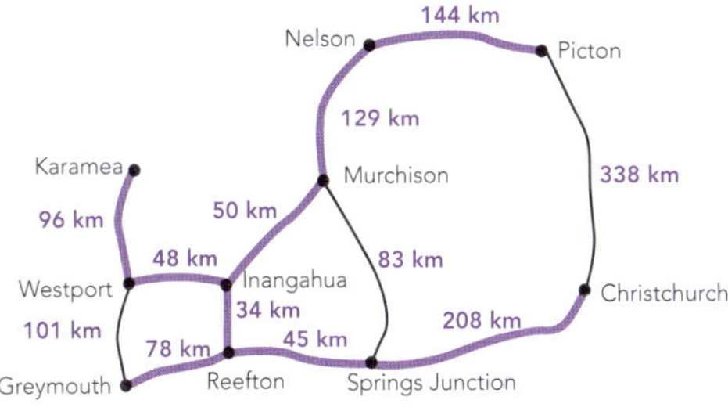

5 Length of minimum spanning tree: 9.6 h

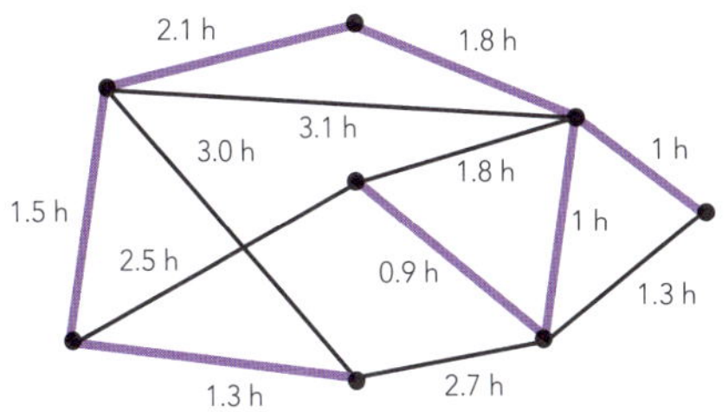

6 Length of minimum spanning tree: 456 km

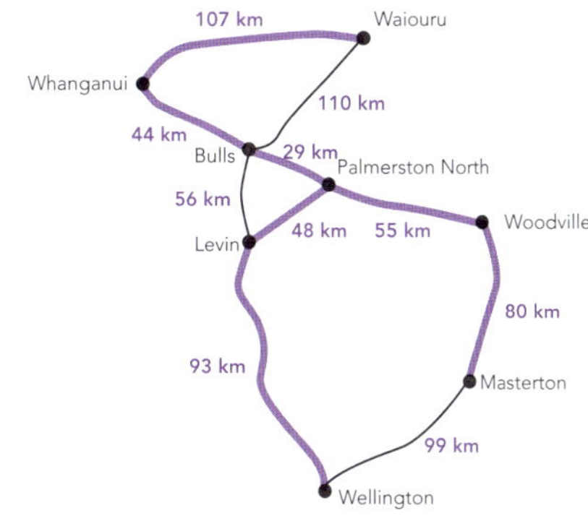

7 Length of minimum spanning tree: 44 943 km

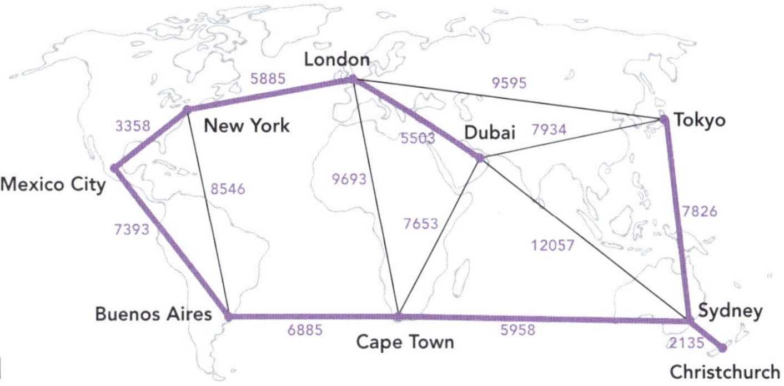

Maximum spanning trees (pp. 47–49)

1 Length of maximum spanning tree: 52 km

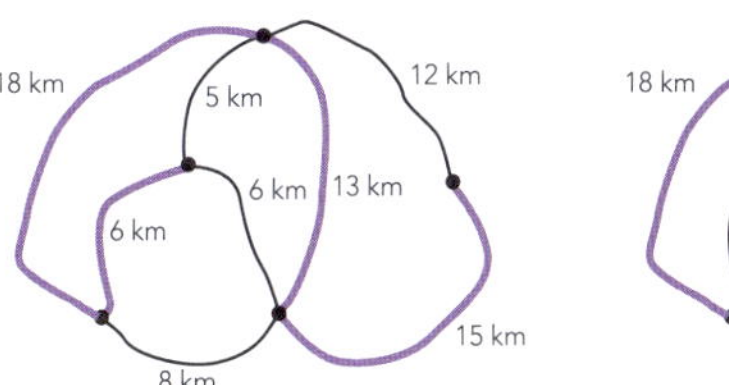

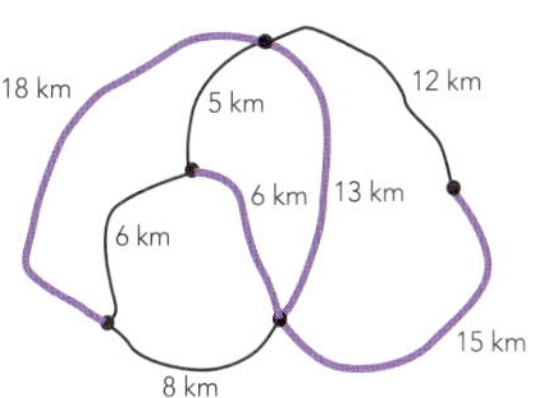

2 Length of maximum spanning tree: 33 minutes

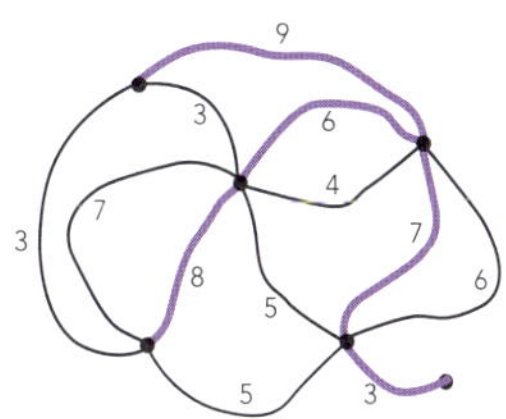

3 Length of maximum spanning tree: 106 nautical miles

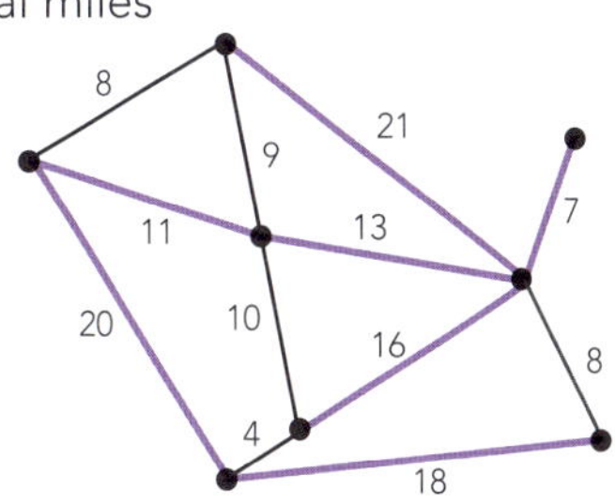

4 Length of maximum spanning tree: 1192 km

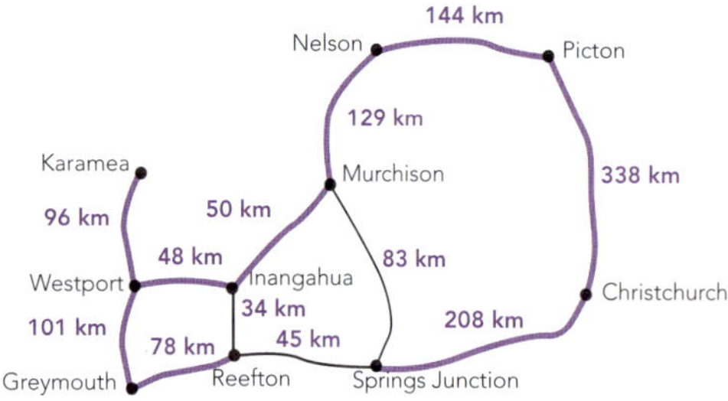

5 Length of maximum spanning tree: 16.5 h

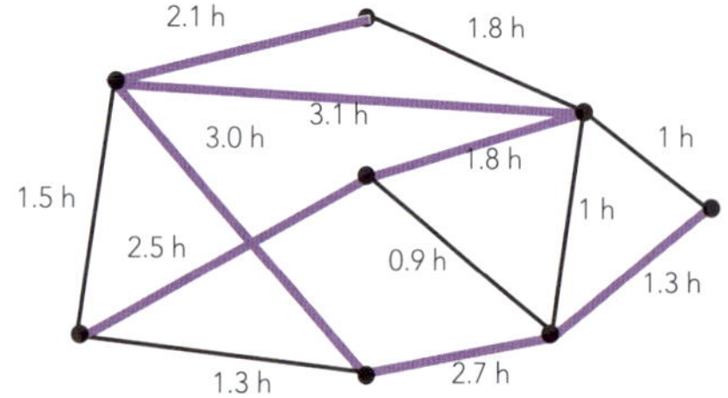

6 Length of maximum spanning tree: 600 km

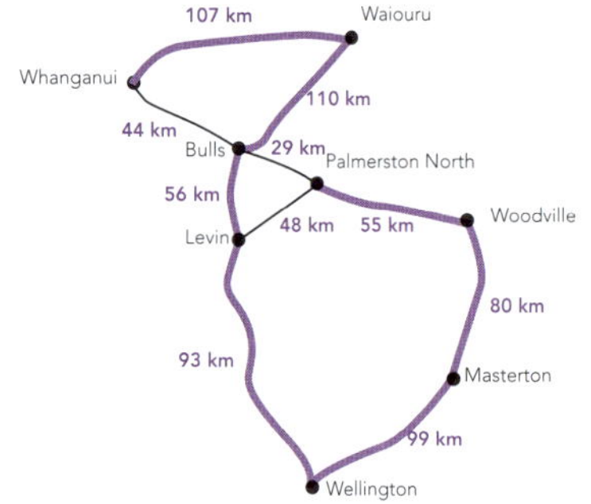

7 Length of maximum spanning tree: 64 238 km

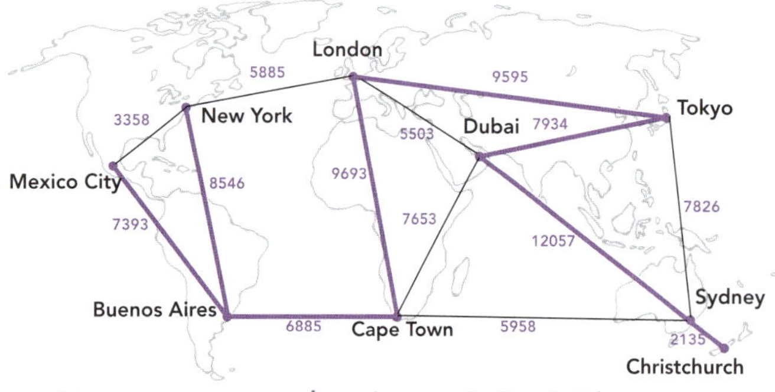

Zed: Maximum spanning tree = 52 rest areas

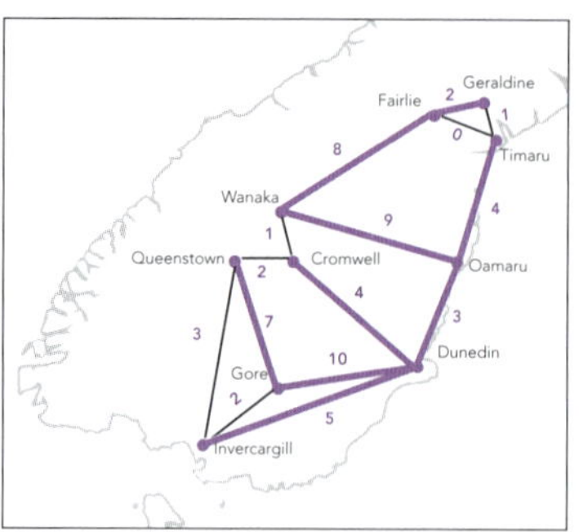

Alfred: Minimum spanning tree = 783 km

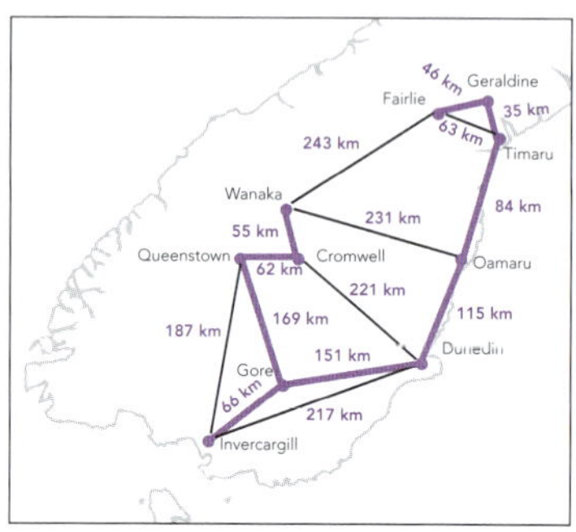

Yelena: Every town except Geraldine and Dunedin is an odd node. Therefore the network is not traversable, which means it is not possible to drive every road exactly once. However, it is possible to visit every town without driving the same road twice, but some roads are missed out. For example, in this case, five roads are not driven:

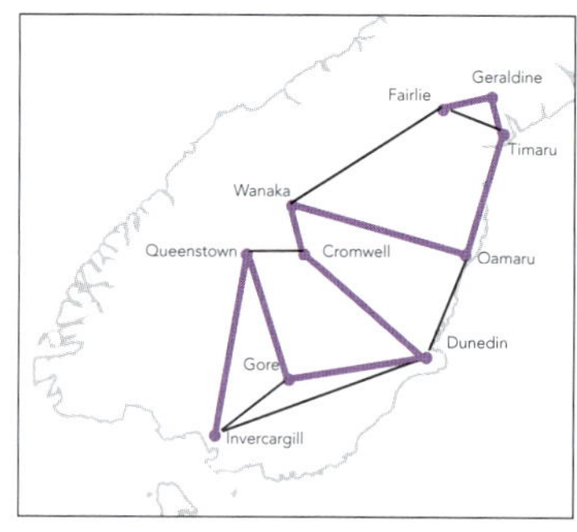

Optimal route: Ask your teacher to check this.

Practice tasks (pp. 50–61)

Practice task one (pp. 50–53)

Xavier: Shortest path = 479 km

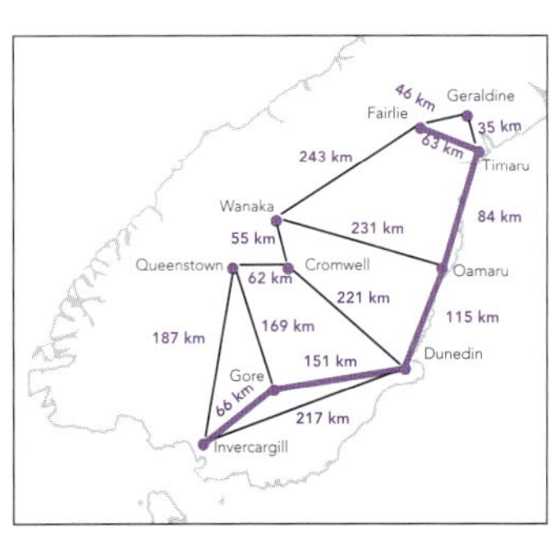

or

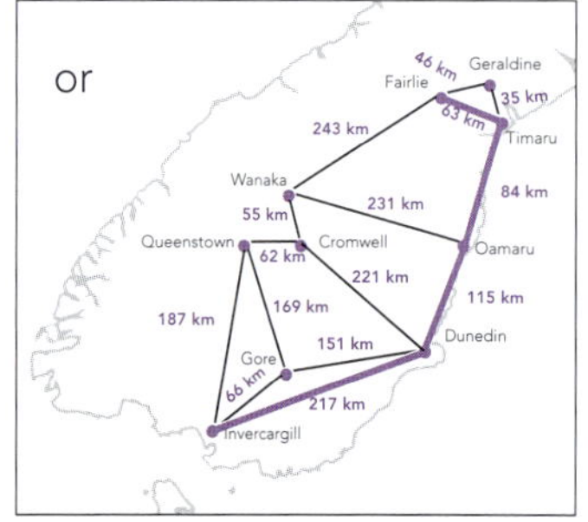

Practice task two (pp. 55–58)

Bella: Shortest route = 427 m

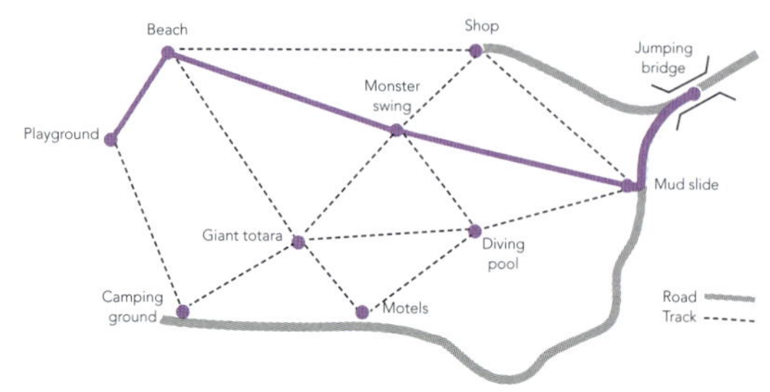

 ISBN: 9780170389433

Aroha: Longest route = 1362 m

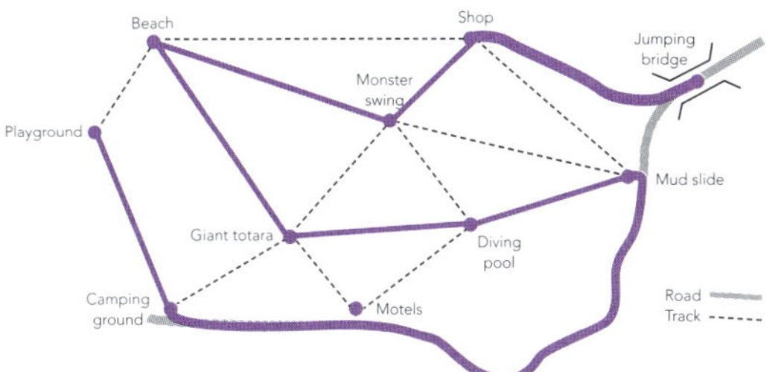

Jack: Removing the tracks to the mud slide and the motels leaves a track network with no odd nodes. This means that it is traversable, and it is possible to construct an Euler circuit.

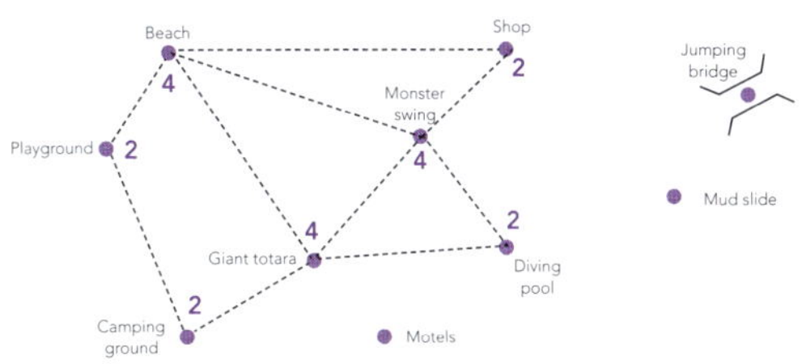

A possible circuit that he could use:
Camping ground → Playground → Beach → Shop → Monster swing → Diving pool → Giant totara → Camping ground

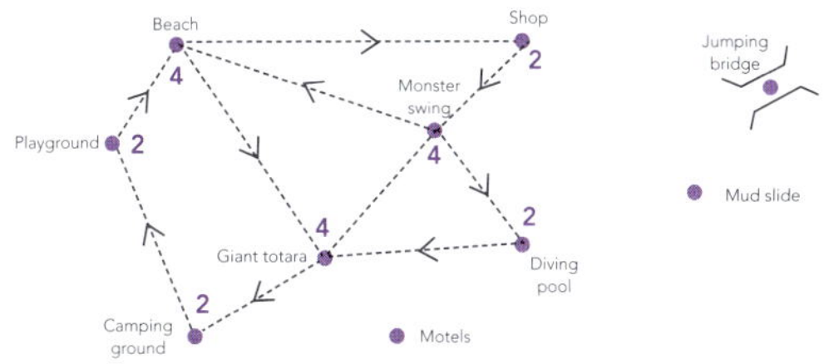

Eli:
Length = 611 m
Cost = 611 x $65 = $39 715

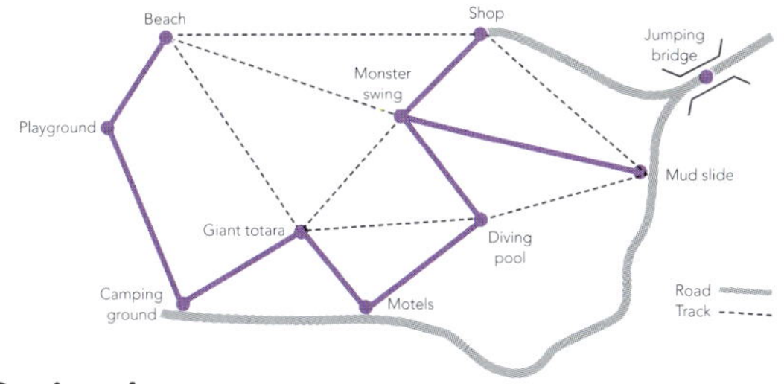

Option A:
Machine-digging:
Length = 611 – 2 x 10 = 591 m
Cost = 591 x $65 = $38 415
Hand-digging:
Length = 38 m
Cost = 38 x $240 = $9120
Total cost = $38 415 + $9120 = $47 535

Option B:
New route:

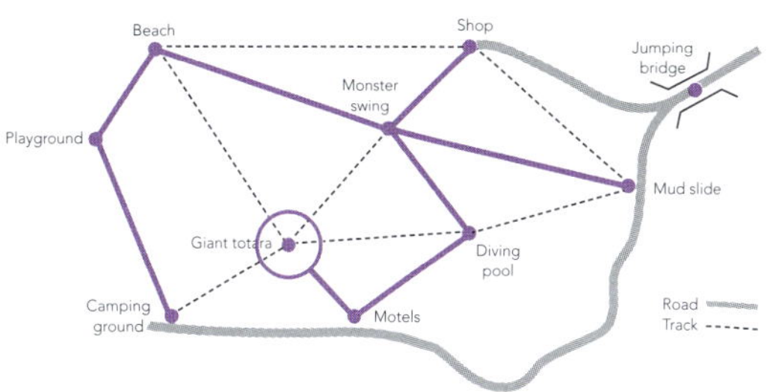

Length = 729 m
Cost = 729 x $65 = $47 385

Recommendation to the council:
The cheapest option is Option B, which costs $150 less than Option A. This is a relatively small amount of money compared with the total cost for installing the lighting (about 0.32%). However, both routes involve long circuits in some circumstances when it is dark. Option A means there is a long route from the monster swing to the beach. Option B results in a long circuit from the giant totara to the camping ground.

Practice task three (pp. 58–61)
Elves' transport of toys from the factory to the wrapping room.
Elves' shortest route: 371 m

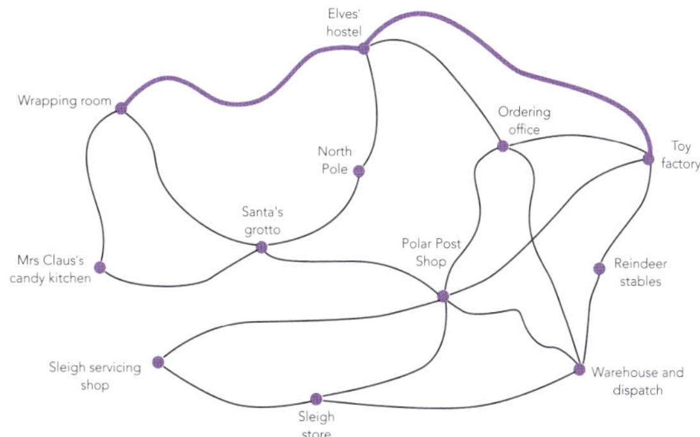

Santa's skiing route
It is possible for Santa to ski every path in the town exactly once because there are exactly two odd nodes: at the wrapping room and the sleigh store. However, because there are two odd nodes, he could not ski a circuit and return to his starting point; this would require that there were no odd nodes. He would need to either start from the wrapping room and finish at the sleigh store, or start at the sleigh store and finish at the wrapping room.

ISBN: 9780170389433

Possible path: Sleigh store → Sleigh servicing shop → Polar Post Shop → Sleigh store → Warehouse and dispatch → Ordering office → Toy factory → Reindeer stables → Warehouse and dispatch → Polar Post Shop → Toy factory → Elves' hostel → Ordering office → Polar Post Shop → Santa's grotto → Mrs Claus's candy kitchen → Wrapping room → Santa's grotto → North Pole → Elves' hostel → Wrapping room.

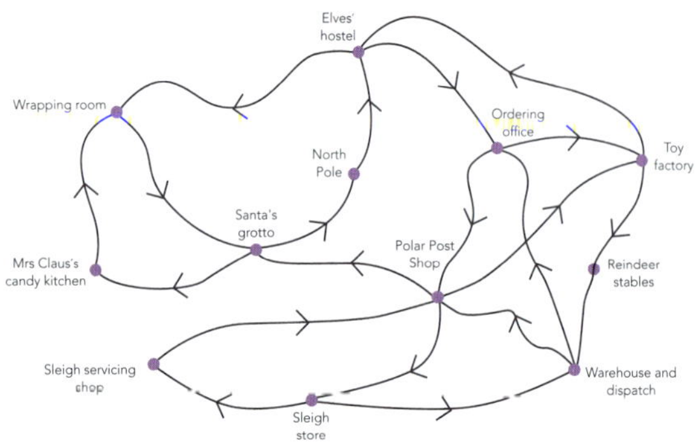

Route A for the monorail

Length of shortest spanning tree: 994 m

Cost = 994 x $450 = $447 300

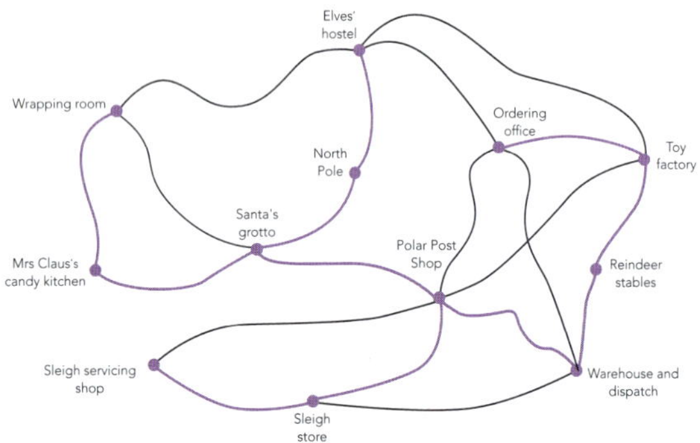

Route B for the monorail

Length of shortest spanning tree, apart from the 39 m: 823 m

Cost = 823 x $450 = $370 350

Difference from Route A: $447 300 – $370 350 = $76 950

∴ Route B becomes cheaper if the 39 m of monorail costs less than $\frac{\$76\,950}{39}$ = $1973.08.

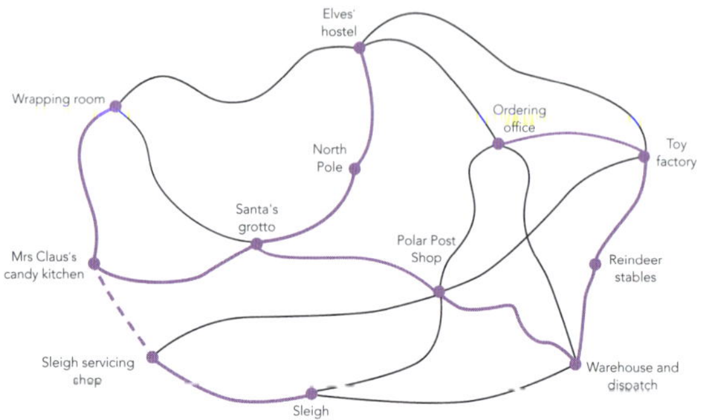

ISBN: 9780170389433